Climate Finance in Africa and Developing Countries

This book explains complex environment of climate finance in Africa, emphasizing the critical need for financial resources to reduce the effects of climate change in the region. It investigates specific obstacles that African countries face when seeking climate funding such as poor financial institutions and lack of technical experience. It offers practical options for closing the climate financing gap, pushing for stronger legislative frameworks, capacity-building programs, and the possible alternative sources.

Key Features

> Addresses timely and relevant issues about existing gaps and future sustainable climate finances in Africa.
>
> Reviews the existing impact of climate change, such as floods, droughts, food security, issues of sustainable development goals, issues of funds, policies, and others.
>
> Includes the strengths and weaknesses of climate change mitigation and adaptation in Africa.
>
> Underlines the importance of increased public and private investment, new financing arrangements, and stronger international cooperation in fulfilling Africa's climate finance needs.
>
> Emphasizes the exploration of innovative financing mechanisms.

This book is aimed at researchers and graduate students in climatology, environmental sciences, policymakers, climate activists, and other stakeholders in climate research.

Climate Finance in Africa and Developing Countries

Bridging the Gap and Building a Sustainable Future

Meseret Dawit Teweldebrihan and
Megersa Olumana Dinka

CRC Press
Taylor & Francis Group
Boca Raton London New York

CRC Press is an imprint of the
Taylor & Francis Group, an **informa** business

First edition published 2026
by CRC Press
2385 NW Executive Center Drive, Suite 320, Boca Raton FL 33431

and by CRC Press
4 Park Square, Milton Park, Abingdon, Oxon, OX14 4RN

CRC Press is an imprint of Taylor & Francis Group, LLC

ISBN: 9781041081241 (hbk)
ISBN: 9781041082248 (pbk)
ISBN: 9781003644378 (ebk)

DOI: 10.1201/9781003644378

Typeset in Times
by Newgen Publishing UK

Contents

Preface

Climate change is no longer a looming threat; it is a pressing issue that requires immediate action. The gravity of the situation is much greater than initially thought. Countries all over the world are working to mitigate the effects of climate change, with climate finance emerging as an important tool. Developing countries, despite contributing minimally to the crisis, bear the brunt of its consequences. Addressing climate change requires a multifaceted strategy. We need more research to figure out how to best mitigate climate change and build resilience. This, in turn, necessitates significant financial investment and a greater global commitment than current efforts. Thus, a collaboration and collective action are critical in addressing this complex problem.

This book provides a more comprehensive understanding and delves into current research on climate finance in Africa, providing a thorough examination of the subject. The book is designed to provide a clear roadmap: the first chapter provides an overview of climate finance and introduces the Africa-specific context. The second chapter delves into the complexities of Africa's climate finance landscape, examining the various challenges and implications. The third chapter examines the current state of African countries, with a focus on their ratification of global environmental agreements. The fourth chapter investigates African countries' capacity to manage climate finance, highlighting potential opportunities and challenges. The fifth chapter contains detailed case studies on climate financing in Africa. Finally, the sixth chapter concludes with valuable recommendations to improve climate finance initiatives across the continent.

Acknowledgements

The authors would like to thank the Seed Grant for New African Principal Investigators (SG-NAPI) award supported by the German Ministry of Education and Research (BMBF) through United Nations Educational, Scientific and Cultural Organization (UNESCO) – The World Academy of Sciences (TWAS) with the financial number (SG-NAPI Agreement) [No. FR3240330995].

About the Authors

Meseret Dawit Teweldebrihan is an advisor, climate, and adaptation expert working in multi-sectoral environments for more than ten years. She is an independent researcher and has taken leadership in different climate relevant projects for more than seven years. Dr. Meseret has solid experience in turning the most challenging experiences into the best achievements of her professional career to benefit her community. She works as a consultant and is involved in different projects. Dr. Meseret's work is mainly on climate change with gender-sensitive, hydrology, irrigation optimization, socio-hydrology, hydraulic engineering, GIS, land and water development, and related fields. Meseret's PhD thesis dissertation titled "Optimizing Irrigation Efficiency of Surface-Groundwater with respect to Climate Change and Gender-Sensitive," and MSc degree in Water Science Engineering (specialization in Hydraulic Engineering Land and Water Development) from the UNESCO-IHE, Delft, the Netherlands, in 2014. She obtained a BSc degree in Soil and Water Engineering and Management from Haramaya University, Ethiopia, in 2011. She has published more than 24 journal articles in accredited publications.

Megersa Olumana Dinka is a graduate with a PhD from the University of Natural Resources and Applied Life Science (Vienna) in 2010. He also did postdoctoral research at Tshwane University of Technology (2012–2014). He has an MSc in Irrigation Engineering from Arba Minch University (Ethiopia). He has more than 18 years of experience as an academician and about 22 years of experience as a researcher. Currently, he is the Full Professor and Head of the Department of Civil Engineering Science at University of Johannesburg. He has expert knowledge in water resource engineering discipline specific to hydrology, hydraulics, and water management aspects. He has taught various courses and modules at undergraduate and postgraduate levels successfully. Currently, he is teaching hydrology, hydraulics, and water treatment technology modules at the University of Johannesburg. Moreover, he also supervised a number of postgraduate students (32 MSc and 10 PhD) successfully. He has published more than 70 journal articles, 2 books, 10 book chapters, and 20 conference proceedings in accredited publications.

Acronyms

AEs	Accredited Entities
AF	Adaptation Fund
AR5	Fifth Assessment Report
CCAP	Climate Change Action Plan
CDM	Clean Development Mechanism
CERF	Climate and Energy Response Facility
CIFs	Climate Investment Funds
CRGE	Climate-Resilient Green Economy
CO2	Carbon dioxide
COP	Conferences of the Parties
DCF	Developed Climate Finance
EPACC	Ethiopian Programme of Adaptation to Climate Change
ETS	Emissions Trading Scheme
EU	European Union
FRLD	Fund for Responding to Loss and Damage
GBT	Green Budget Tagging
GCF	Green Climate Fund
GDP	Gross Domestic Product
GEF	Global Environment Facility
GFCF	Gross Fixed Capital Formation
GHG	Greenhouse Gas
GIS	Geographic Information Systems
INDC	Intended Nationally Determined Contribution
LDCs	Least Developed Countries
LDF	Loss and Damage Fund
LLA	Locally Led Adaptation
LTF	Long-Term Climate Finance
LT-LEDS	Long-Term Low Emission Development Strategy
LTMS	Long-Term Mitigation Scenario Process
LUREFET	Law for the Use of Renewable Energies and for the Finance of the Energy Transition
MEFCC	Ministry of Environment, Forest, and Climate Change
MoPD	Ministry of Planning and Development
NAPA	National Adaptation Programme of Action
NAPCC	National Action Plan on Climate Change

NAP-ETH	Ethiopian National Adaption Plan
NCCC	National Council for Climate Change
NCQG	New Collective Quantified Goal
NDA	National Designated Authority
NDC	Nationally Determined Contributions
OECD	Organisation for Economic Co-operation and Development
PPPs	Public-Private Partnerships
SDGs	Sustainable Development Goals
UNFCCC	United Nations Framework Convention on Climate Change
USA	United States of America

Introduction to the Climate Change and Climate Financing

1

PREFACE

This chapter discusses climate change and its impact on Africa and developing countries, focusing on how the continent is disproportionately affected despite its minor contribution. Climate finance, or funding for mitigation (lowering emissions) and adaptation (building resilience), is portrayed as an important tool. However, there is a significant gap between the funds required and what is currently available. The document highlights ways to close the gap, such as improved policies, capacity building, and investigating alternative funding sources. Overall, successfully addressing climate change in Africa necessitates a multi-faceted international effort.

1.1 INTRODUCTION TO CLIMATE FINANCE IN THE GLOBE

Climate finance has evolved as a crucial global imperative, diverting financial resources toward mitigating and adapting to the growing difficulties of climate change (Songwe et al., 2022). This concept has quickly evolved from a minor

DOI:10.1201/9781003644378-1

worry to a major pillar of world policy to facilitate the transition to a low-carbon, climate-resilient economy. Climate finance aims to transform the global financial industry into a proactive engine of sustainable development by directing investments toward clean energy, encouraging carbon trading systems, and fostering innovation and broader economic reforms (Khan & Munira, 2021; Songwe et al., 2022). Climate financing is urgent because of climate change's huge environmental difficulties, necessitating an immediate and coordinated global response (Boumis et al., 2023; Khan & Munira, 2021). The financial sector is critical to channeling investments toward long-term objectives. The Kyoto Protocol sets the framework for incorporating financial mechanisms into climate policy, allowing institutions such as banks and pension funds to play an active role in the low-carbon transition. However, maximizing the potential of financial markets needs a more thorough integration of financial strategies into climate policy talks. Ethical considerations are central to the climate finance debate, including questions about equality, responsibility, and allocating duties and rewards (Gordon, 2023). These moral issues must be thoroughly addressed to guarantee a just and equitable global response to climate change that is consistent with the goals of the United Nations Framework Convention on Climate Change (UNFCCC). Fairly dividing the costs and benefits of climate action is critical for building international cooperation and guaranteeing the long-term viability of climate financing initiatives (Khan & Munira, 2021; Songwe et al., 2022). Nations are increasingly prioritizing climate change through legislative action, despite definitional challenges, with organizations such as GLOBE International facilitating critical cross-party dialogues to develop effective national strategies for emissions reduction, resilience enhancement, and stronger international negotiation stances.

Countries like Japan pioneered climate legislation, but the most substantial measures arose after 2008, possibly influenced by the 2009 UNFCCC negotiations. Ongoing legislative efforts in China, Mexico, and South Africa reflect a long-term commitment to combating climate change (Table 1.1). Many countries have passed "flagship" legislation combining existing and new climate policies into complete frameworks (Table 1.1). Five five-year plans play a similar integrative role in China and India, demonstrating a comprehensive approach to climate governance.

1.1.1 The Global Landscape of Climate Vulnerability and Climate Finance Options

The global landscape of climate vulnerability and finance is a complex patchwork of varied national interests and degrees of susceptibility to climate change (Buchner et al., 2014; Burton et al., 2006). Historical contingencies, such as Russia's Kyoto Protocol benefit from surplus emission credits during an economic downturn, have a significant impact on financial flows, whereas examples such as Indonesia's forestry legislative achievements, facilitated by Norwegian funding, demonstrate the transformative potential of external finance on domestic policy (Avdeeva, 2005; Cochran & Pauthier, 2019). In contrast, countries like the United States prioritize energy security, tying climate legislation to lowering foreign oil dependence through local energy development (Cochran & Pauthier, 2019). National leadership, often in tandem with hosting major international events, drives the timing of climate legislation, as evidenced by Japan, Indonesia, Mexico, and South Africa's strategic use of international venues to promote domestic agendas, as well as the UK's 2008 Climate Change Act following the Gleneagles G8 Summit (Lorenzoni & Benson, 2014). However, contextual circumstances such as elections, as witnessed with Canada's legislative delays, can stymie development. The EU's unique position enables it to use climate leadership to promote internal cohesion and global influence. Developing countries, particularly those that are very sensitive to climate change, such as South Africa and India, prioritize adaptation by including it in national planning, recognizing its crucial impact on key sectors (Table 1.1). While affluent countries such as the United Kingdom recognize the necessity of adaptation, mitigation remains their primary priority, emphasizing nations' various objectives and capacities in addressing the numerous challenges of climate change (Table 1.1).

1.2 INTRODUCTION TO THE CLIMATE FINANCE IN AFRICA AND DEVELOPING COUNTRIES

Climate financing in Africa and developing countries faces a stark reality of unfairness, with those least responsible for climate change bearing the brunt of the repercussions (Gordon, 2023; Nor, 2025). These regions, which have unique geographical insights critical for future climate projections, urgently require a strong and equitable financial framework that empowers them to develop solutions. Although the UNFCCC's Annex I/Non-Annex I classification and commitments such as the $100 billion annual commitment attempted to address this imbalance, the actual allocation and usefulness of these payments are still debated (Nakhooda et al., 2011; Ogede et al., 2024). Climate-related disasters cause considerable economic losses in sub-Saharan

TABLE 1.1   The laws and the climate change flagship legislation

NO.	YEAR	COUNTRY	DETAILS AND DESCRIPTION
1	1998 (modified 2005)	Japan	The "Law Concerning the Promotion of Measures to Cope with Global Warming" in Japan, enacted in 1998 and amended in 2005, established a multi-tiered legal framework for addressing climate change by integrating national and local efforts.
2	2005 and 2008	Mexico	Mexico demonstrated its commitment to climate action by establishing the Inter-Secretariat Commission on Climate Change in 2005 and 2008 to harmonize national climate policies, as well as reinforcing its commitment to the LUREFET, which promotes renewable energy integration and clean technology deployment through financial tools and strategic plans, with the goal of diversifying the energy matrix and reducing reliance on fossil fuels.
3	2007	Italy	In 2007, Italy's Climate Change Action Plan (CCAP) was a significant legislative effort to meet the Kyoto Protocol's GHG reduction targets, establishing a framework for national climate change mitigation strategies and policies and showcasing Italy's commitment to global sustainability.
4	2007	Canada	In 2007, Canada's Kyoto Protocol Implementation Act required the government to develop and implement measures to reduce greenhouse gas emissions, ensuring the country met its Kyoto Protocol responsibilities and demonstrated its commitment to global climate action.
5	2007 (amended 2008)	Germany	Germany's Integrated Climate and Energy Programme, launched in 2007 and refined in 2008, represented a significant step forward in environmental sustainability, aiming to reduce greenhouse gas emissions by 40% from 1990 levels by 2020 through a comprehensive strategy that focused on the building sector but was later expanded to include targeted actions for transportation and construction.

TABLE 1.1   (Continued)

NO.	YEAR	COUNTRY	DETAILS AND DESCRIPTION
6	2007 (updated in 2008 and 2009)	China	China's National Climate Change Programme, launched in 2007 and updated in 2009, represented a significant strategic shift by incorporating emissions reduction, adaptation, technological advancement, public awareness, and institutional capacity-building into a comprehensive framework designed to accelerate renewable energy adoption and mitigate greenhouse gas emissions through specific policy measures such as strengthened energy laws and increased renewable energy deployment.
7	2008	European Union	European Union's 2008 Climate and Energy Package (CARE) established a strong legal framework that included Emissions Trading Scheme (ETS), reform, national effort-sharing, renewable energy promotion, & CO_2 storage regulations, all aimed at aggressively mitigating climate change & driving the adoption of sustainable energy across the EU.
8	2008	India	In 2008, India launched its National Action Plan on Climate Change (NAPCC), a comprehensive policy framework outlining the country's strategy for mitigating and adapting to climate change through eight focused "national missions" targeting key sectors such as solar energy and water resources, with the initial goal of guiding climate initiatives until 2017 and establishing a sustainable development path.
9	2008	Indonesia	Indonesia established a comprehensive climate change framework in 2008 with the Presidential Regulation on the National Council for Climate Change (NCCC), a high-level body chaired by the President. It comprised 17 ministers who centralized and coordinated national climate policymaking, supported by specialized working units addressing key areas such as adaptation, emissions reduction, and forestry.

(continued)

TABLE 1.1   (Continued)

NO.	YEAR	COUNTRY	DETAILS AND DESCRIPTION
10	2008	South Africa	South Africa's "Vision, Strategic Direction, and Framework for Climate Policy," developed through extensive public engagement and the LTMS, outlined a comprehensive climate action plan in 2008 that included ambitious greenhouse gas reductions, enhanced initiatives, transformative "Business Unusual" actions, adaptation strategies, and stakeholder partnerships, paving the way for the 2012 draft "Zero" Climate Change Policy.
11	2008	United Kingdom	The Climate Change Act of 2008 in the United Kingdom is a watershed moment in legislative history because it establishes a comprehensive framework with legally binding targets, such as an 80% reduction in emissions from 1990 levels by 2050 (later enhanced) and five-year carbon budgets, to drive the transition to a low-carbon economy through strategic carbon management and sustainable technology investment, while also fostering innovation and economic opportunities.
12	2009	United States of America	In 2009, despite the absence of comprehensive federal climate legislation, the U.S.A government advanced climate action through Executive Order 13514, which imposed strict GHG emission reduction mandates on federal agencies, and the American Recovery and Reinvestment Act, which allocated substantial funds to stimulate renewable energy and energy efficiency initiatives.
13	2009	South Korea	South Korea demonstrated its commitment to sustainability in 2009 by enacting the Framework Act on Low Carbon Green Growth, a landmark law that integrated environmental considerations into its economic strategy by establishing GHG emission reduction targets, instituting a cap-and-trade mechanism, and encouraging renewable energy adoption, resulting in a comprehensive framework for climate change mitigation & environmentally responsible economic progress.

TABLE 1.1   (Continued)

NO.	YEAR	COUNTRY	DETAILS AND DESCRIPTION
14	2009	Russia	In 2009, Russia established the Climate Doctrine, a legal framework that outlined the country's climate policy strategy, emphasizing increased scientific research, mitigation and adaptation measures, and active international collaboration to address the multi-faceted challenges of climate change.
15	2009	Brazil	Brazil strengthened its commitment to climate action in 2009 by establishing the National Policy on Climate Change (NPCC), which incorporated previous initiatives such as the National Plan on Climate Change and the National Fund on Climate Change, thereby strengthening its strategy for emissions reduction and climate adaptation in accordance with UNFCCC obligations.
16	2009 and 2020	France	France's Grenelle I and II legislation, enacted in 2009 and 2010, establish a comprehensive legislative framework for achieving national sustainability through specific environmental goals such as emission reductions, renewable energy growth, and increased energy efficiency.

Africa, which is particularly vulnerable because of its reliance on rain-fed agriculture and insufficient adaptive capacity. Despite receiving global and bilateral climate finance, the distribution within the area is unequal, with fragile and conflict-affected nations underserved, and the aggregate amount is insufficient to fulfill the enormous demands (Mugambiwa & Kwakwa, 2022; Nakhooda et al., 2011). A donor-centric approach, which frequently ignores the specific requirements and goals of key recipient countries such as South Africa, Nigeria, Kenya, Tanzania, Ethiopia, Mozambique, Niger, and Senegal, impedes the efficiency of climate finance. Addressing fundamental policy problems about fund allocation criteria and the impact of recipient characteristics is critical to ensuring that climate finance helps African countries build resilience and achieve sustainable development.

The effectiveness of climate finance is further impeded by a donor-centric approach that frequently fails to address recipient nations' specific needs and priorities. This disparity emphasizes the importance of donors and receivers having a common understanding of the goals and expectations of climate funding. Key policy problems arise: What criteria do donor countries employ to allocate climate financing to African countries? Do unique recipient characteristics influence their financing decisions? Addressing these issues is critical for increasing the impact and equity of climate finance, ensuring that it effectively supports African countries' attempts to build resilience and achieve sustainable development in the face of climate change.

1.2.1 Overview of Climate Change and Climate Finance

Globally, the climate varies naturally over long periods of time but is now experiencing faster and major adjustments known as "climate change," which is characterized by long-term changes in weather patterns and event distributions (Timilsina, 2021). Climate variability has historically been driven by natural processes such as volcanic eruptions and solar oscillations (Stenchikov, 2021). However, contemporary scientific agreement relates the past century's rapid climate change to anthropogenic activity. The extensive use of fossil fuels and other human-caused activities has resulted in a significant increase in atmospheric GHG concentrations, particularly CO_2 (Yoro & Daramola, 2020). These gases, by trapping solar energy, cause global temperatures to gradually rise, a phenomenon known as global warming. Human-caused global warming is the primary cause of climate change's diverse and negative impacts.

Despite Africa's small contribution to global emissions, climate change remains a major concern. Africa has numerous difficulties, including severe droughts and desertification, unpredictable weather patterns, and increasing sea levels (Bannor, 2022; Barnard et al., 2014; Green et al., 2011). These worries jeopardize essential infrastructure, water supply, food security, and the continent's general economic progress. While Africa's vulnerability is obvious, its resilience and adaptive capabilities are multi-faceted and diversified. The continent's ability to cope and adapt to climate change is typically constrained by financial constraints, technological gaps, and institutional restrictions (Conway et al., 2015; Misra, 2014). However, some local governments and communities are implementing novel adaptation tactics, such as drought-resistant crop varieties, enhanced water management practices, and ecosystem-based adaptation. Furthermore, indigenous knowledge is vital in promoting resilience since it provides critical insights into dealing with environmental variability. Despite these efforts, Africa's ability to cope with the growing effects of climate change and assure a sustainable future will require significant investment in climate-resilient infrastructure, capacity building, and climate finance.

Climate finance has emerged as an important tool in response to these challenges. It refers to the financial resources for climate change initiatives, including adaptation and mitigation measures (Kellogg, 2019; Kløve et al., 2014). Adaptation strategies, which include building more resilient infrastructure or developing drought-resistant crops, seek to assist communities in adapting to the unavoidable consequences of climate change (Brown et al., 2010; Nelson et al., 2009; Timilsina, 2021). Mitigation strategies, on the other hand, aim to reduce greenhouse gas emissions and prevent future global warming. By funding these initiatives, climate financing can help African countries build resilience, transition to a low-carbon future, and protect their development prospects.

1.2.2 Global Warming/Climate Change: An Increase Threat

Climate change, primarily caused by human activity over the last century, is an increasingly serious worldwide issue, fundamentally altering the statistical distribution of long-term weather patterns (Jones, 1999; Salinger, 2005; Teweldebrihan & Dinka, 2024). This change is manifested by an increase in extreme weather events, such as enhanced storms, floods, droughts, and heatwaves, as well as increased climatic variability, which hampers forecasting and planning. Projections show that average sea levels and extreme weather events will climb significantly by 2050, accompanied by highly irregular rainfall patterns, with some regions seeing increased precipitation and others experiencing severe droughts (Boumis et al., 2023; Kirezci et al., 2020; Vousdoukas et al., 2018). Furthermore, the expected global temperature increase of more than 2°C, as described in the 2015 Paris Agreement, has far-reaching repercussions, including melting glaciers, rising sea levels, and changed agricultural output (Betts et al., 2018; Coffel et al., 2017; Hoegh Guldberg et al., 2018). As a result, immediate and persistent efforts are required to counteract these effects by reducing greenhouse gas emissions, investing in renewable energy, and widespread adoption of sustainable habits, so as to ensure a more robust and sustainable future.

1.2.2.1 Climate change: a global crisis

Climate change is a global catastrophe marked by dramatic changes in Earth's long-term weather patterns, mostly caused by human activity during the last two centuries (Betts et al., 2018; Green et al., 2011). The release of massive amounts of greenhouse gases such as carbon dioxide and methane from fossil fuel combustion and deforestation has resulted in an unprecedented warming trend, with each of the last four decades exceeding the previous in heat, culminating in the 2011–2020 decade being the hottest on record (Nica et al., 2019; Yoro & Daramola, 2020). The interconnected effects of global warming include more frequent and intense extreme weather events, rising sea levels due to melting ice, ocean acidification, and potential disruptions to essential resources such as food and water, all of which pose serious threats to ecosystems and human well-being (Gahlawat & Lakra, 2020). Understanding the complex causes and far-reaching implications of climate change is critical for developing effective mitigation and adaptation policies that ensure a sustainable future (Figure 1.1).

1.2.2.2 The causes of climate change

Different studies indicate that human activities are the primary cause of climate change (Coffel et al., 2017). Carbon dioxide and other greenhouse gases are released into the atmosphere when power is generated, goods are manufactured, and forests are cut down (Ritchie & Roser, 2017; Yoro & Daramola, 2020). Consequently, most scholars reveal that transportation, food production, and building power generation all contribute to these emissions (Carlin, 2006; Nayak et al., 2020; Wang & Azam, 2024). On the other hand, overconsumption and resource waste exacerbate climate change (Driga & Drigas, 2019).

The Earth's climate has naturally changed over history, undergoing periods of both cooling and warming, which are mostly determined by the delicate energy balance between land, sea, and atmosphere (Wang & Azam, 2024; Yoro & Daramola, 2020). However, the current rapid warming trend is mostly due to human activity (Table 1.2), which has greatly increased the concentration GHG in the atmosphere. The combustion of fossil fuels such as coal, oil, and natural gas emits large volumes of carbon dioxide that have been trapped for millions of years (Wang & Azam, 2024). Similarly, deforestation, whether for timber or agricultural development, depletes key carbon sinks and releases stored carbon into the atmosphere, worsening the warming impact. These activities break the natural balance, trapping heat that would otherwise be radiated back into space, gradually causing the planet to overheat. This enhanced greenhouse impact has far-reaching implications. Projections predict that world average temperatures could rise by 1.8–4.0°C by 2100 (Coffel et al., 2017; Nayak et al., 2020). Such a temperature increase will cause enormous environmental changes, including the rapid melting of polar ice caps and glaciers, resulting in a large rise in sea levels. This directly threatens low-lying coastal areas, potentially displacing people and destroying essential infrastructure.

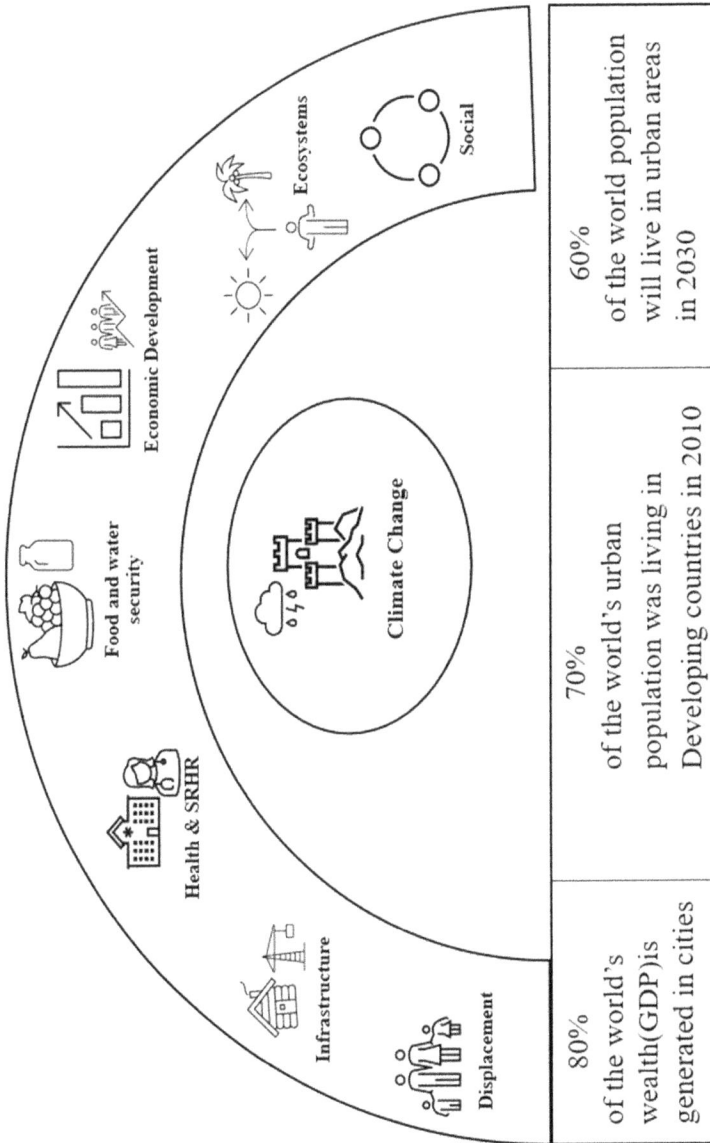

FIGURE 1.1 The causes and consequences of climate change. (Elaborated by the author.)

TABLE 1.2   The causes and effects of global climate change

CATEGORY	CAUSE (HUMAN ACTIVITIES)	EFFECT (CONSEQUENCES)
Primary Driver	• Rising levels of greenhouse gases (GHGs) in the atmosphere.	• Rapid and unprecedented changes in the global climate.
Energy Production	• Fossil fuels (coal, oil, natural gas) are used for electricity generation.	• The release of massive amounts of carbon dioxide (CO_2).
Manufacturing	• Energy-intensive industrial operations (extraction & production).	• Increased greenhouse gas emissions from factories and supply chains.
Deforestation	• Rampant cutting down of forests.	• Elimination of CO_2-absorbing trees and release of stored carbon.
Transportation	• Reliance on fossil fuel-powered vehicles (cars, trucks, planes, ships).	• Significant contribution to greenhouse gas emissions.
Food Production	• Livestock farming (methane), fertilizer use (nitrous oxide).	• Emission of greenhouse gases, including methane (CH_4) and nitrous oxide (N_2O).
Building Operations	• Uses fossil fuels for heating, cooling, and lighting.	• Increased carbon footprint from building energy use.
Consumption Patterns	• Overconsumption of goods and services.	• Higher emissions due to increasing production, transportation & waste.
Temperature	• Increased greenhouse gas emissions.	• Hotter temperatures, heatwaves, wildfires, melting polar ice caps.
Extreme Weather	• Warming trend.	• Severe storms with greater severity and frequency.
Water Resources	• Increased temperatures.	• Increased drought, water scarcity, crop failures, and desertification.
Ocean Impacts	• Thermal expansion and melting ice.	• Rising sea levels, coastal flooding, erosion, and disruption of marine ecosystems.
Biodiversity	• Rapid climate changes.	• Loss of species, habitat alteration, and difficulty for animals to adapt.

TABLE 1.2   (Continued)

CATEGORY	CAUSE (HUMAN ACTIVITIES)	EFFECT (CONSEQUENCES)
Food Security	• Extreme weather, drought, and ecosystem disruption.	• Food shortages, hunger, malnutrition.
Human Health	• The increase in temperatures and pollution.	• Heatstroke, the spread of infectious infections, and respiratory issues.
Socioeconomic Impacts	• Resource scarcity, displacement, economic instability.	• Increased displacement, leading to a circle of vulnerability.

Furthermore, climate change is anticipated to alter precipitation patterns, resulting in more frequent and severe extreme weather events like floods and droughts. Dry places are expected to grow drier, while other areas may receive more rainfall, upsetting agricultural methods and water resources (Nayak et al., 2020). Increased temperatures impact weather systems, altering wind patterns, precipitation, and the frequency of severe weather events, such as hurricanes and storms (Betts et al., 2018; Wang & Azam, 2024).

Human activities are the primary cause of this increase in GHG emissions, with carbon dioxide being the most significant source (Table 1.2). The combustion of fossil fuels, a cornerstone of modern industrial society, emits massive amounts of CO2 into the atmosphere. Deforestation, another important issue, not only destroys carbon sinks but also releases stored carbon when trees are burned. Furthermore, agricultural activities contribute to GHG emissions, as cattle produce methane via enteric fermentation, and nitrogen-based fertilizers emit nitrous oxide. The rapid growth of the world population, together with the rising demand for energy and resources, has accelerated these activities, resulting in a steady increase in GHG emissions (Wang & Azam, 2024). This increase in emissions is expected to have serious consequences, particularly in vulnerable places such as Africa. According to predictions, millions of Africans will face severe water stress due to climate change by 2020, and rain-fed agriculture yields may fall by up to 50% in some areas. By 2080, dry and semi-arid territory in Africa is expected to increase, worsening food insecurity. Climate change is also projected to influence the distribution of infectious diseases, possibly increasing the prevalence of mosquito-borne disorders. Rising sea levels also represent a substantial threat to coastal communities that face environmental issues such as erosion and floods.

1.2.2.3 Implications of climate change

Climate change has far-reaching and profound consequences for the environment, health, economy, and society (Betts et al., 2018). Rising global temperatures exacerbate extreme weather events such as stronger storms, prolonged droughts, and severe heatwaves, which directly affect vulnerable populations and ecosystems (Driga & Drigas, 2019; Wang & Azam, 2024). Ocean warming and sea-level rise endanger coastal people, while biodiversity faces enormous threats from habitat loss and shifting climatic patterns (Mollema et al., 2015; Nelson et al., 2010). Food security is challenged by dwindling crop yields and limited water supplies.

Furthermore, climate change heightens health hazards by increasing the prevalence of heat-related illnesses and accelerating the spread of infectious diseases (Table 1.2). Aside from these direct effects, the economic ramifications are enormous, necessitating the implementation of carbon pricing policies to reduce greenhouse gas emissions; however, political difficulties and differing estimates of marginal impacts provide significant challenges (Nica et al., 2019; Tracy, 2024). The wide range of effects, from agricultural shifts to changes in energy demand and sea-level rise, highlights the importance of comprehensive assessment and aggregate indicators in guiding policy decisions and public understanding, ensuring a holistic approach to addressing this complex global crisis.

1.2.3 Building Resilience through Climate Finance

Climate change poses a complex challenge requiring a multi-faceted approach (Barnard et al., 2014). Financial resources are critical in addressing the global crisis, where climate finance comes in (Buchner et al., 2014). It is the lifeblood of efforts to combat climate change and build resilience, particularly in developing countries.

Climate finance has two main components: mitigation and adaptation. Mitigation focuses on reducing greenhouse gas emissions, the root cause of climate change (Brown et al., 2010; Buchner et al., 2014). This entails allocating funds to clean energy solutions such as renewable energy sources and energy-saving technologies. By encouraging the development and adoption of these alternatives, we can significantly reduce our reliance on fossil fuels and slow the rate of global warming. The other important aspect is adaptation. Hence, climate change's effects are already felt worldwide, but developing countries often lack the resources to cope. Climate finance intervenes to provide critical assistance for adaptation efforts (Brown et al., 2010; Chirambo, 2017). This includes investments in early warning systems for extreme weather events,

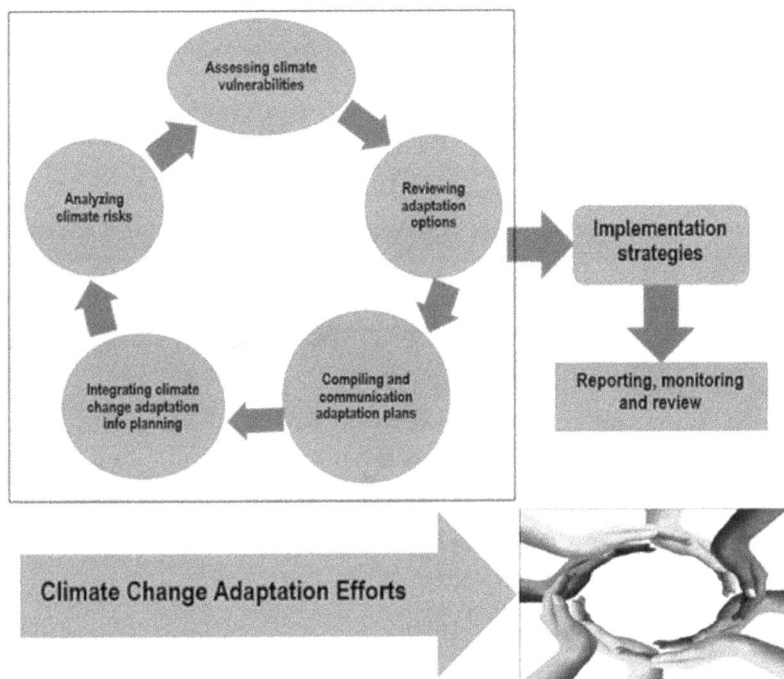

FIGURE 1.2   Climate change adaptation plan process. (Elaborated by the author.)

climate-smart agricultural practices that can adapt to changing weather patterns, and the creation of resilient infrastructure capable of enduring floods, droughts, and rising sea levels (Chirambo, 2014; Timilsina, 2021). Climate finance assists vulnerable communities in building resilience and weathering the unavoidable storms of climate change by strengthening adaptation capabilities (Figure 1.2).

Climate finance is a strategic investment in a more sustainable future for all (Arezki, 2021; Brown et al., 2010). It enables developing countries to participate actively in the fight against climate change (Figure 1.3) while encouraging innovation and paving the way for a greener, more resilient global community (Chirambo, 2014).

1.2.4 Africa's Climate Finance Challenge: A Funding Gap Threatens Progress

While Africa bears the brunt of climate change's devastating consequences, its historical contribution to greenhouse gas emissions is negligible (Chirambo,

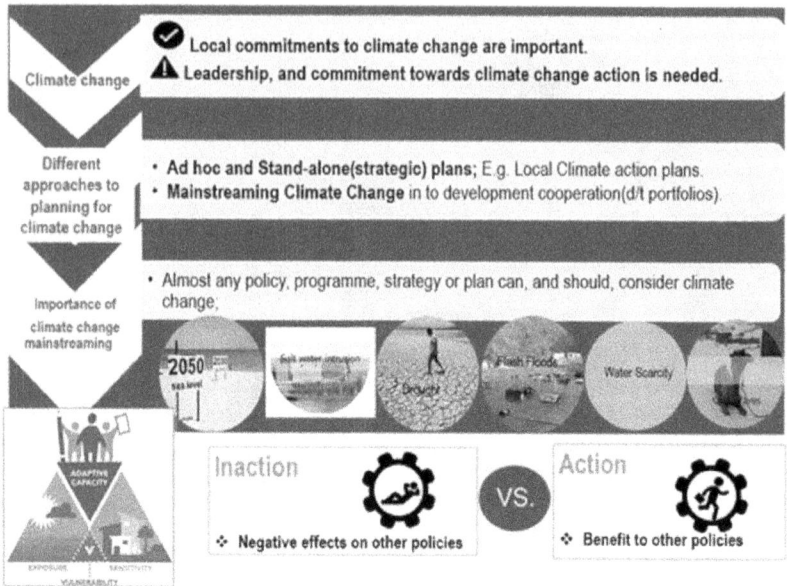

FIGURE 1.3 Climate change vulnerability and adaptive capacity. (Elaborated by the author.)

2017). Consequently, African countries have outlined their commitments to climate action through Nationally Determined Contributions (NDCs) under the Paris Agreement (Buchner et al., 2014; Cochran & Pauthier, 2019). These plans detail critical steps for mitigation, which reduces emissions, and adaptation, which increases resilience to climate impacts (Chirambo, 2014). However, implementing these plans depends on one crucial factor access to adequate funding.

Unfortunately, Africa's current financial landscape for climate action paints a bleak picture. According to estimates, implementing the continent's NDCs will cost between $1.3 trillion and $2.8 trillion annually (Buchner et al., 2014; Timilsina, 2021). This enormous need is set against a sobering reality: Africa currently receives only $30 billion in annual climate finance flows. This massive funding gap impedes progress, limiting African countries' ability to reduce emissions, adapt to changing weather patterns, and build long-term resilience.

Historically, Africa's climate finance has been primarily funded by public sources such as bilateral aid from developed countries and multilateral development banks. While this support is critical, it falls far short of the continent's requirements (Chirambo, 2014; Tamasiga et al., 2023; Timilsina, 2021). As a result, there is an increasing push to leverage private sector investment.

Unlocking this potential could be a game changer, boosting Africa's climate action efforts (Arezki, 2021).

1.2.5 Challenges Hinder Africa's Access to Climate Finance

While climate finance provides a critical lifeline for Africa in combating climate change, mobilizing these resources presents significant challenges (Arezki, 2021). One major impediment is limited financial access (Barnard et al., 2014). Many African countries suffer from weak financial institutions and systems, making it difficult to meet the stringent requirements and complex procedures commonly associated with international climate funds. This translates to a situation in which high-potential projects fail due to a lack of access to critical funding.

Furthermore, the perception of high investment risks deters private sector participation in climate change projects. African countries are frequently regarded as risky investments due to political unrest or a lack of established infrastructure (Chirambo, 2014; Tamasiga et al., 2023; Timilsina, 2021). This reluctance from private entities reduces the overall capital available for climate action (Figure 1.4).

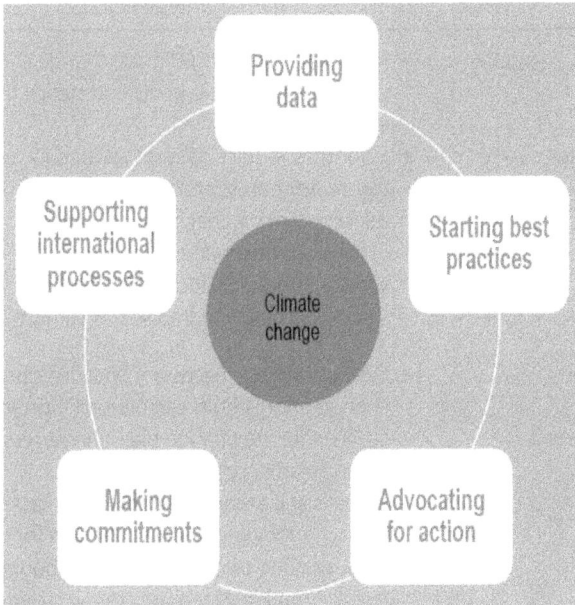

FIGURE 1.4   The cyclic action on climate change. (Elaborated by the author.)

The frequent lack of institutional capacity and technical expertise in African countries is compounding these issues. Successfully managing climate finance projects necessitates strong governance structures, skilled personnel, and knowledge of financial management best practices. Without these elements, even if funding becomes available, it won't be easy to use it effectively to meet climate goals (Table 1.3).

1.2.5.1 Bridging the gap: solutions for effective climate finance mobilization

A multi-faceted approach is required to address these challenges and maximize the potential of climate finance in Africa (Arezki, 2021; Martín Casas & Remalia Sanogo, 2022). First and foremost, efforts must be directed toward improving policy frameworks. Streamlining approval processes, establishing clear guidelines, and providing incentives for climate-friendly investments can make a more appealing environment for both domestic and international actors (Martín Casas & Remalia Sanogo, 2022; Tamasiga et al., 2023). De-risking mechanisms, which provide guarantees or insurance against potential losses, can also help to increase private sector participation. Second, building capacity

TABLE 1.3   The climate change adaptation principle checklist

CATEGORY	SYMBOL	DETAIL DESCRIPTION AND PROCESS
Target		• Plans for combating climate change should contain quantifiable goals and clearly defined outcomes.
Requirements		• Countries require data on existing circumstances, future estimates, and projected repercussions to address climate change successfully.
Strategies		• Strong climate change planning will necessitate various strategies, including policies that address the broad-reaching impacts of climate change.
Collaboration		• Collaboration is necessary because climate change affects both sectors and jurisdictional scales. Climate change planning necessitates widespread support.
planning		• There is still a significant implementation gap in planning for climate change. Plans for mitigating and adapting to climate change abound, but very few are implemented, tracked, or even provide a clear process.

is critical. This includes investing in training programs, technical assistance, and knowledge-sharing initiatives to provide African institutions with the skills and expertise required to manage climate finance effectively. Building strong institutions promotes transparency, accountability, and efficient project implementation, making these countries more appealing destinations for climate investment.

Finally, looking into alternative funding options can help diversify the resource pool. Innovative mechanisms such as green bonds, which raise capital for climate projects, and debt-for-nature swaps, which forgive a portion of a country's debt in exchange for conservation efforts, provide promising alternatives to traditional financing channels (Chirambo, 2014; Martín Casas & Remalia Sanogo, 2022). By exploring these opportunities, African countries can progress toward a climate-resilient future.

1.2.6 The Urgency of Bridging the Climate Finance Gap in Africa

Africa is particularly vulnerable to the effects of climate change despite having the lowest contribution to global greenhouse gas emissions (Arezki, 2021; Kissinger et al., 2019). This harsh reality highlights the critical need for climate finance or financial resources provided to developing countries to combat climate change.

Closing the climate finance gap is critical to Africa's long-term development and resilience (Barnard et al., 2014; Mekonnen, 2014; Tamasiga et al., 2023). Estimates indicate a significant gap between the funds required and those currently available (Martín Casas & Remalia Sanogo, 2022; Mekonnen, 2014). This lack of resources limits Africa's ability to carry out critical climate action plans.

A three-pronged approach is critical to ensure Africa has the tools to build a resilient, low-carbon future (Arezki, 2021; Mekonnen, 2014). First and foremost, greater international cooperation is required (Chirambo, 2017; Savvidou et al., 2021). Developed nations must keep their promises to provide financial assistance to developing countries. This includes increasing the amount of climate finance allocated to Africa and improving access to these funds (Baskaran et al., 2023; Kissinger et al., 2019).

Second, it is critical to investigate alternative financing options (Bannor, 2022; Mungai et al., 2021). Carbon markets, green bonds, and blended finance are all examples of innovative approaches that can open up new capital sources (Arezki, 2021; Conway & Schipper, 2011). By diversifying funding sources, Africa can reduce its reliance on traditional aid while attracting private sector investment in climate-friendly solutions (Chirambo, 2017; Conway et al., 2015).

Finally, a strong focus on capacity development is required (Arezki, 2021; Digitemie & Ekemezie, 2024; Mungai et al., 2021). Providing African countries with the technical expertise and infrastructure to effectively manage and use climate finance will ensure its long-term impact. This includes training local experts, strengthening institutions, and developing knowledge-sharing networks across the continent (Belianska et al., 2022; Chirambo, 2017). The international community can help Africa bridge the climate finance gap by implementing these solutions (Buchner et al., 2014; Chaudhury, 2020; Dinku et al., 2014; Park, 2022). This will pave the way for the continent's more sustainable and resilient future, eventually benefiting the entire planet.

1.3 CONCLUSION

In conclusion, despite Africa's minor role in climate change, the continent bears the consequences. Climate finance is a critical tool for addressing these challenges. It includes financial resources for mitigation, which reduces emissions, and adaptation, which increases resilience. However, there is a significant gap between the funding required and what is currently available. Several obstacles impede Africa's access to climate finance, including limited financial resources, high-risk perceptions, and a lack of institutional capacity. Thus, to close this gap, a multi-faceted approach is required. The critical steps are improving policy frameworks, launching capacity building initiatives, and investigating alternative funding options. The importance of bridging the climate financing gap cannot be overstated. It is vital for Africa's long-term development and resilience in the face of climate change. Africa can empower itself to build a sustainable and resilient future by implementing solutions that involve increased international cooperation, exploring alternative financing, and focusing on capacity development. This, in turn, benefits the entire world.

The report emphasizes the importance of climate finance in solving the rising global climate issue, particularly in vulnerable places such as Africa. Climate financing is critical for reducing greenhouse gas emissions and adjusting to the inevitable effects of climate change. However, there is a large financial vacuum, particularly in Africa, where the need for climate action far outstrips the available resources. Thus, it focuses on the ethical implications of climate financing, emphasizing the importance of equitable cost and reward distribution, particularly for developing countries that face the brunt of climate change despite making minor contributions to it. It also highlights the importance of national legislation and international collaboration in pushing climate action, citing examples of countries that have pioneered climate laws and participated in global forums.

The report delves deeper into Africa's access to climate finance barriers, such as inadequate financial access, perceived high investment risks, and insufficient institutional capacity. It argues for a multimodal approach to closing the gap, including better legislative frameworks, increased capacity, and exploring alternative funding structures such as green bonds and debt-for-nature swaps. Finally, the statement emphasizes eliminating the climate funding gap to promote sustainable development and resilience in vulnerable regions. It advocates for stronger international cooperation, expanded private sector participation, and creative financing options to guarantee that climate finance successfully supports global efforts to mitigate climate change.

REFERENCES

Arezki, R. (2021). Climate finance for Africa requires overcoming bottlenecks in domestic capacity. *Nature Climate Change*, *11*(11), 888–888.

Avdeeva, T. G. (2005). Russia and the Kyoto Protocol: Challenges ahead. *Review of European, Comparative & International Environmental Law*, *14*, 293.

Bannor, F. (2022). *Agriculture, climate change and technical efficiency: The case of sub-Saharan Africa.* University of Johannesburg.

Barnard, S., Watson, C., Greenhill, R., Caravani, A., Trujillo, N. C., Hedger, M., & Whitley, S. (2014). *Climate finance: Is it making a difference?* Overseas Development Institute.

Baskaran, G., Ekeruche, A., Heitzig, C., Ordu, A. U., & Senbet, L. W. (2023). Financing climate-resilient infrastructure in Africa.

Belianska, A., Bohme, N., Cai, K., Diallo, Y., Jain, S., Melina, M. G., … Zerbo, S. (2022). *Climate change and select financial instruments: An overview of opportunities and challenges for Sub-Saharan Africa.* International Monetary Fund.

Betts, R. A., Alfieri, L., Bradshaw, C., Caesar, J., Feyen, L., Friedlingstein, P., … Morfopoulos, C. (2018). Changes in climate extremes, fresh water availability and vulnerability to food insecurity projected at 1.5 C and 2 C global warming with a higher-resolution global climate model. *Philosophical Transactions of the Royal Society A: Mathematical, Physical and Engineering Sciences*, *376*(2119), 20160452.

Boumis, G., Moftakhari, H. R., & Moradkhani, H. (2023). Coevolution of extreme sea levels and sea-level rise under global warming. *Earth's Future*, *11*(7), e2023EF003649.

Brown, J., Cantore, N., & te Velde, D. W. (2010). *Climate financing and development.* Overseas Development Institute.

Buchner, B., Stadelmann, M., Wilkinson, J., Mazza, F., Rosenberg, A., & Abramskiehn, D. (2014). Global landscape of climate finance 2019. *Climate Policy Initiative*, *32*(1), 1–38.

Burton, I., Diringer, E., & Smith, J. (2006). *Adaptation to climate change: International policy options*. Citeseer.

Carlin, A. (2006). Global climate change control: Is there a better strategy than reducing greenhouse gas emissions. *University of Pennsylvania Law Review*, *155*, 1401.

Chaudhury, A. (2020). Role of intermediaries in shaping climate finance in developing countries—lessons from the green climate fund. *Sustainability*, *12*(14), 5507.

Chirambo, D. (2014). The climate finance and energy investment dilemma in Africa: Lacking amidst plenty. *Journal of Developing Societies*, *30*(4), 415–440.

Chirambo, D. (2017). Enhancing climate change resilience through microfinance: Redefining the climate finance paradigm to promote inclusive growth in Africa. *Journal of Developing Societies*, *33*(1), 150–173.

Cochran, I., & Pauthier, A. (2019). *A framework for alignment with the Paris Agreement: Why, what and how for financial institutions* (Vol. 56). Institute for Climate Economics.

Coffel, E. D., Horton, R. M., & De Sherbinin, A. (2017). Temperature and humidity based projections of a rapid rise in global heat stress exposure during the 21st century. *Environmental Research Letters*, *13*(1), 014001.

Conway, D., & Schipper, E. L. F. (2011). Adaptation to climate change in Africa: Challenges and opportunities identified from Ethiopia. *Global Environmental Change*, *21*(1), 227–237.

Conway, D., Van Garderen, E. A., Deryng, D., Dorling, S., Krueger, T., Landman, W., … Ringler, C. (2015). Climate and southern Africa's water–energy–food nexus. *Nature Climate Change*, *5*(9), 837–846.

Digitemie, W. N., & Ekemezie, I. O. (2024). Assessing the role of climate finance in supporting developing nations: A comprehensive review. *Finance & Accounting Research Journal*, *6*(3), 408–420.

Dinku, T., Block, P., Sharoff, J., Hailemariam, K., Osgood, D., del Corral, J., … Thomson, M. C. (2014). Bridging critical gaps in climate services and applications in Africa. *Earth Perspectives*, *1*, 1–13.

Driga, A. M., & Drigas, A. S. (2019). Climate change 101: How everyday activities contribute to the ever-growing issue. *International Journal of Recent Contributions from Engineering, Science & IT*, *7*(1), 22–31.

Gahlawat, I. N., & Lakra, P. (2020). Global climate change and its effects. *Integrated Journal of Social Sciences*, *7*(1), 14–23.

Gordon, N. J. (2023). Climate finance: An overview. *Environment: Science and Policy for Sustainable Development*, *65*(4), 18–26.

Green, T. R., Taniguchi, M., Kooi, H., Gurdak, J. J., Allen, D. M., Hiscock, K. M., … Aureli, A. (2011). Beneath the surface of global change: Impacts of climate change on groundwater. *Journal of Hydrology*, *405*(3), 532–560.

Hoegh Guldberg, O., Jacob, D., Taylor, M., Bindi, M., Brown, S., Camilloni, I. A., … Engelbrecht, F. (2018). *Impacts of 1.5 C global warming on natural and human systems*. Intergovernmental Panel on Climate Change.

Jones, J. (1999). Climate change and sustainable water resources: Placing the threat of global warming in perspective. *Hydrological Sciences Journal*, *44*(4), 541–557.

Kellogg, W. W. (2019). *Climate change and society: Consequences of increasing atmospheric carbon dioxide*: Routledge.

Khan, M. R., & Munira, S. (2021). Climate change adaptation as a global public good: Implications for financing. *Climatic Change*, *167*(3), 50.

Kirezci, E., Young, I. R., Ranasinghe, R., Muis, S., Nicholls, R. J., Lincke, D., & Hinkel, J. (2020). Projections of global-scale extreme sea levels and resulting episodic coastal flooding over the 21st century. *Scientific Reports*, *10*(1), 1–12.

Kissinger, G., Gupta, A., Mulder, I., & Unterstell, N. (2019). Climate financing needs in the land sector under the Paris Agreement: An assessment of developing country perspectives. *Land Use Policy*, *83*, 256–269.

Kløve, B., Ala-Aho, P., Bertrand, G., Gurdak, J. J., Kupfersberger, H., Kværner, J., … Rossi, P. M. (2014). Climate change impacts on groundwater and dependent ecosystems. *Journal of Hydrology*, *518*, 250–266.

Lorenzoni, I., & Benson, D. (2014). Radical institutional change in environmental governance: Explaining the origins of the UK Climate Change Act 2008 through discursive and streams perspectives. *Global Environmental Change*, *29*, 10–21.

Martín Casas, N., & Remalia Sanogo, A. (2022). Climate Finance in West Africa: Assessing the state of climate finance in one of the world's regions worst hit by the climate crisis.

Mekonnen, A. (2014). Economic costs of climate change and climate finance with a focus on Africa. *Journal of African Economies*, *23*(suppl_2), ii50–ii82.

Misra, A. K. (2014). Climate change and challenges of water and food security. *International Journal of Sustainable Built Environment*, *3*(1), 153–165.

Mollema, P. N., Antonellini, M., Dinelli, E., Greggio, N., & Stuyfzand, P. J. (2015). The influence of flow-through saline gravel pit lakes on the hydrologic budget and hydrochemistry of a Mediterranean drainage basin. *Limnology and Oceanography*, *60*(6), 2009–2025.

Mugambiwa, S., & Kwakwa, M. (2022). Multilateral climate change financing in the developing world: Challenges and opportunities for Africa. *International Journal of Research in Business and Social Science (2147–4478)*, *11*(9), 306–312.

Mungai, E. M., Ndiritu, S. W., & Da Silva, I. (2021). Unlocking climate finance potential for climate adaptation: Case of climate smart agricultural financing in sub Saharan Africa. In *African handbook of climate change adaptation* (pp. 2063–2083). Springer.

Nakhooda, S., Caravani, A., Bird, N., Schalatek, L., & America, H. (2011). *Climate finance in sub-Saharan Africa*. Climate Finance Policy Briefs, Heinrich Böll Stiftung North America, Washington, DC, USA and Overseas Development Institute (ODI).

Nayak, H., Yadav, S. P., & Yadav, D. K. (2020). Contribution of natural and anthropogenic activities in greenhouse gases emission. *Energy*, *4*(2), 1–4.

Nelson, G. C., Rosegrant, M. W., Koo, J., Robertson, R., Sulser, T., Zhu, T., … Batka, M. (2009). *Climate change: Impact on agriculture and costs of adaptation* (Vol. 21). International Food Policy Research Institute.

Nelson, G. C., Rosegrant, M. W., Palazzo, A., Gray, I., Ingersoll, C., Robertson, R., …
Ringler, C. (2010). *Food security, farming, and climate change to 2050: scenarios,
results, policy options* (Vol. 172). International Food Policy Research Institute.

Nica, A., Popescu, A., & Ibanescu, D.-C. (2019). Human influence on the climate
system. *Current Trends in Natural Sciences*, *8*(15), 209–215.

Nor, M. I. (2025). Investigating the dynamics of climate finance disbursements: A panel
data approach from 2003 to 2022. *PloS One*, *20*(3), e0318170.

Ogede, J. S., Oduola, M. O., & Tiamiyu, H. O. (2024). Income inequality and carbon
dioxide (CO_2) in sub-Saharan Africa countries: The moderating role of financial
inclusion and institutional quality. *Environment, Development and Sustainability*,
26(7), 18385–18409.

Park, J. (2022). How can we pay for it all? Understanding the global challenge of
financing climate change and sustainable development solutions. *Journal of
Environmental Studies and Sciences*, *12*(1), 91–99.

Ritchie, H., & Roser, M. (2017). *CO_2 and other greenhouse gas emissions*. Our world
in data.

Salinger, M. J. (2005). Climate variability and change: Past, present and future–an overview. *Climatic Change*, *70*(1), 9–29.

Savvidou, G., Atteridge, A., Omari-Motsumi, K., & Trisos, C. H. (2021). Quantifying
international public finance for climate change adaptation in Africa. *Climate
Policy*, *21*(8), 1020–1036.

Songwe, V., Stern, N., & Bhattacharya, A. (2022). *Finance for climate action: Scaling
up investment for climate and development*. Grantham Research Institute on
Climate Change and the Environment, London School of Economics and Political
Science.

Stenchikov, G. (2021). The role of volcanic activity in climate and global changes. In
Climate Change (pp. 607–643). Elsevier.

Tamasiga, P., Molala, M., Bakwena, M., Nkoutchou, H., & Onyeaka, H. (2023). Is
Africa left behind in the global climate finance architecture: Redefining climate
vulnerability and revamping the climate finance landscape – A comprehensive
review. *Sustainability*, *15*(17), 13036.

Teweldebrihan, M., & Dinka, M. (2024). The impact of climate change on the development of water resources. *Global Journal of Environmental Science and
Management*. *10*(3), 1359–13705.

Timilsina, G. R. (2021). Financing climate change adaptation: International initiatives.
Sustainability, *13*(12), 6515.

Tracy, S. J. (2024). *Qualitative research methods: Collecting evidence, crafting analysis, communicating impact*. John Wiley & Sons.

Vousdoukas, M. I., Mentaschi, L., Voukouvalas, E., Verlaan, M., Jevrejeva, S., Jackson,
L. P., & Feyen, L. (2018). Global probabilistic projections of extreme sea levels
show intensification of coastal flood hazard. *Nature Communications*, *9*(1), 2360.

Wang, J., & Azam, W. (2024). Natural resource scarcity, fossil fuel energy consump-
tion, and total greenhouse gas emissions in top emitting countries. *Geoscience
Frontiers*, *15*(2), 101757.

Yoro, K. O., & Daramola, M. O. (2020). CO_2 emission sources, greenhouse gases, and
the global warming effect. In *Advances in carbon capture* (pp. 3–28). Elsevier.

The Landscape of Climate Financing in Africa

2

2.1 INTRODUCTION

Africa, although contributing minimally to global carbon emissions, bears a disproportionate share of the terrible effects of climate change (Arezki, 2021; Boumis et al., 2023). The disparity toward climate financing has been highlighted in studies by several scholars (Arezki, 2021; Barnard et al., 2014; Nakhooda et al., 2011). This dramatic gap is worsened by a lack of climate financing, a vital instrument for strengthening resilience to droughts, deserts, severe weather, floods, and rising sea levels (Adenle et al., 2017; Doku et al., 2021; Nakhooda et al., 2011). Climate finance, which includes both mitigation and adaptation initiatives as defined by scholars and stakeholders, is critical for establishing resilience (Adenle et al., 2017; Arezki, 2021; Barnard et al., 2014; Nakhooda et al., 2011). It includes financial resources made available to developing countries to combat climate change through mitigation and adaptation strategies (Chirambo, 2016). However, present and forecast budget levels are insufficient to meet the growing issues of food insecurity, unemployment, and other climate-related concerns (Chirambo, 2016; Mekonnen, 2014).

The importance of addressing climate change has grown significantly in international discourse, as indicated by the increased emphasis on environmental well-being in climate policy and critical assessments undertaken at Conferences of the Parties (COPs). The 2023 Global Stocktake emphasized the significance of including health issues into climate action, acknowledging that combating climate change is inextricably tied to saving lives (Boumis et al., 2023; Chirambo, 2014). Effectively implementing the Paris Agreement 2015,

DOI:10.1201/9781003644378-2

which focuses on decreasing air pollution and controlling global warming, represents a critical chance to enhance global public health and avert countless climate-related deaths. Thus, to effectively estimate climate finance flows, they must be viewed within the larger framework of global investments and funding. While indicators such as gross domestic product (GDP) and gross fixed capital formation (GFCF) shed light on economic activity and investment flows, they do not directly track progress toward sustainable development goals (SDGs) or account for environmental deterioration (Chirambo, 2016; Mekonnen, 2014). Despite increased global climate finance flows, substantial obstacles remain, such as definitional ambiguity, data shortages, and the persistent dominance of fossil fuel funding (Chirambo, 2017; Nakhooda et al., 2011; Tamasiga et al., 2023). The uneven distribution of climate finance, with a few large economies dominating and the most vulnerable countries receiving a disproportionately small amount, emphasizes the need for a more equitable and effective approach.

The formation of the Loss and Damage Fund (LDF) represents a watershed moment in recognizing and tackling climate change's catastrophic effects on vulnerable countries, notably in Africa (Salimi Turkamani, 2024). However, current funding levels fall far short of projected demands, emphasizing the critical need for additional financial contributions from rich countries. While recent climate conferences have resulted in significant pledges for global climate action, funding for the LDF remains insufficient, highlighting the ongoing disparity between mitigation efforts and support for those already suffering the effects of climate change. Addressing this disparity and providing equal access to climate finance are critical for strengthening resilience and promoting sustainable development in Africa and other vulnerable areas.

2.2 CLIMATE VULNERABILITY

The importance of a clean, healthy planet is gaining traction in international debates and policymaking (Barnard et al., 2014; Hidayati, 2021). Thus, the growing recognition of climate vulnerability highlights the turning point in global policy, shifting from environmental preservation to a focus on direct human health consequences (Boumis et al., 2023). International conversations, particularly at COPs and the landmark 2023 Global Stocktake, are cementing agreements to incorporate health issues into climate action, particularly in Nationally Determined Contributions (Barnard et al., 2014; Hidayati, 2021). This shift recognizes that addressing climate change, as specified in the Paris Agreement, is fundamentally a public health responsibility (Table 2.1). Reducing fossil fuel emissions combats both global warming and air pollution,

TABLE 2.1 The climate-vulnerable countries and their health impacts

THE REGION/ COUNTRY TYPE	SOME SPECIFIC EXAMPLES	PRIMARY CLIMATE-RELATED HEALTH CONSEQUENCES	CONTRIBUTING FACTORS
Small Islands Developing States (SIDS)	• Maldives, • Tuvalu, and • Kiribati	• Rising sea levels cause displacement, • Water contamination, and • Increased risk of mental and infectious diseases.	• Low elevation, • Limited resources, • Reliant on coastal habitats.
Coastal Regions in Developing Countries	• Bangladesh, • Vietnam, • Nigeria (coastal areas)	• Flooding, displacement, • Waterborne diseases, • Increased salinity of drinking water and related	• High population, • Low-lying areas, • Inadequate infrastructure.
Arid and Semi-Arid Regions	• Sahel region (e.g., Niger, Chad), • Parts of India, • Part of Australia	• Drought, food insecurity, malnutrition, heat stress, increased risk of infectious diseases.	• Water scarcity, • Desertification, • Temperature increase.
Low-Income Countries with Weak Health Systems	• Sub-Saharan Africa (general), • Parts of Southeast Asia	• Spread of vector-borne diseases (malaria, dengue) • Diarrheal diseases, and • Heat stress.	• Limited access to healthcare, • Sanitation, and • Clean water.
Mountainous Regions	• Nepal, • Peru (Andes)	• Glacial melt leading to water scarcity and floods • Risk of landslides • Change in disease vectors.	• Glacial retreat, • Altered precipitation patterns.
Arctic Regions	• Northern • Canada, • Russia, and • Greenland	• Food insecurity and health impacts • Increased risk of infectious diseases.	• Permafrost thaw, • Sea ice loss, • Ecosystem disruption.

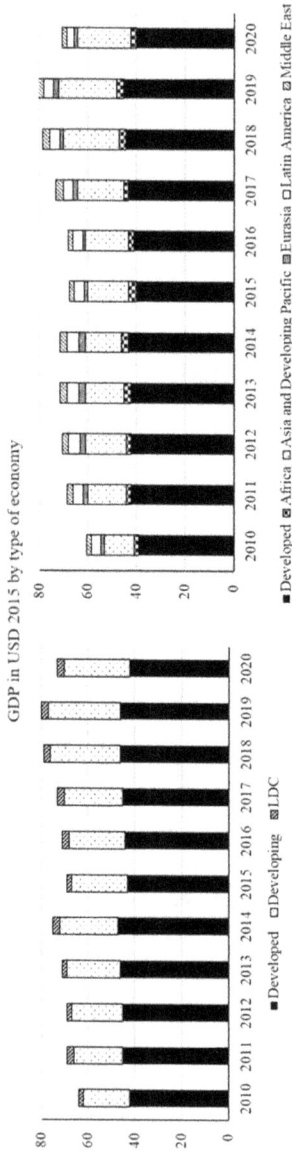

FIGURE 2.1 Financial flows – GDP (trillion USD2015) by type of economy (left) and region (right). (Source: IPCC report chapter 15.)

potentially sparing millions of premature deaths each year (Table 2.1). Furthermore, limiting temperature rise can prevent an estimated 250,000 climate-related deaths per year by mid-century, owing to climate-induced disruptions to food systems, disease proliferation, and intensified extreme weather events, all of which exacerbate conditions such as malnutrition, heat stress, and diarrheal illnesses (Belianska et al., 2022; Kissinger et al., 2019). In essence, climate action is increasingly recognized as an important strategy for protecting human lives and well-being (Chirisa et al., 2021).

2.3 FINANCIAL FLOWS FOR CLIMATE ACTION: A BROADER PERSPECTIVE

To accurately assess financial flows dedicated to climate action, they need to be considered in the context of all investments and financing (Belianska et al., 2022). We can gain a comprehensive understanding of trends over time by examining both flows and stocks simultaneously. Financial flows, measured per unit of time, help to accumulate stocks, which represent the total value of assets at a given point in time (Belianska et al., 2022).

GDP is a key metric for analyzing economic activity. It measures the monetary value of final goods and services produced within a country over a given time period (IPCC et al., 2007). In 2020, global GDP exceeded USD 70 trillion, with developed countries accounting for roughly 60% of the total (Figure 2.1). While GDP provides information about economic activity, it does not directly measure human well-being or progress toward SDGs. It can also positively count activities that have a negative impact on the environment while not taking into account natural resource depletion and degradation.

GFCF is another relevant metric (IPCC & New York, 2007). GFCF includes both tangible assets, such as infrastructure and equipment, and intangible assets. It is a proxy for real-economy investment flows (IPCC & New York, 2007). Globally, GFCF increased by more than 40% between 2010 and 2019, reaching roughly USD 20 trillion in 2019 (Figure 2.2).

2.3.1 Challenges in Estimating Climate Finance Flows

Estimating climate financing flows is a difficult task, hampered by definitional uncertainties, variable reporting standards, and coverage gaps that prevent accurate

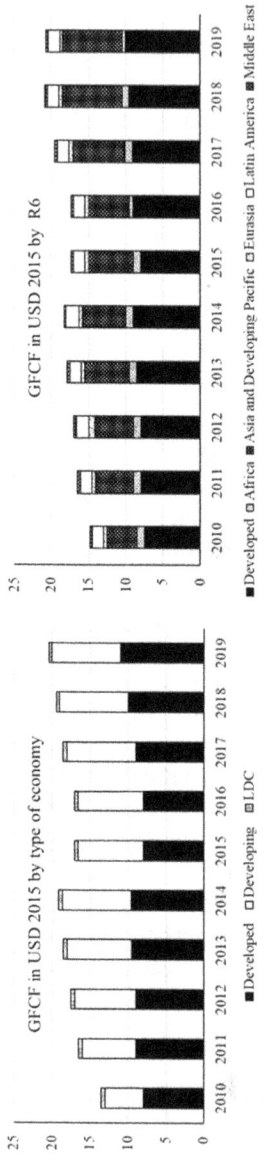

FIGURE 2.2 Financial flows – GFCF (trillion USD2015) by type of economy (left) and region (right). (Source: IPCC report chapter 15.)

evaluations (Belianska et al., 2022; Boumis et al., 2023). While global climate financing has increased since the Fifth Assessment Report (AR5), hitting a high of over USD681 billion in 2016, data reliability remains an issue, with swings and probable decreases reported in later years (Rossitto, 2021; Shirai, 2023). Notably, these transactions account only a modest portion of global financial flows, emphasizing the critical need for significant scaling (IPCC & New York, 2007). In the African context, these challenges are exacerbated by issues such as African leaders' varying willingness and readiness to prioritize climate action, widespread corruption that can divert funds, and the risk of climate funds being misused for political agendas rather than their intended purpose (Belianska et al., 2022). Furthermore, the ongoing flow of funds into fossil fuel projects highlights the challenge of moving away from carbon-intensive investments, hampering efforts to precisely track and effectively use climate finance (Figure 2.3).

Domestic Dominance: Recent research has found that domestic or national sources of climate finance dominate in both developed and developing countries. While cross-border financing is important, it accounts for a small portion of the overall climate finance landscape (Belianska et al., 2022; Shirai, 2023). This highlights the critical role of national policies and settings in shaping climate finance flows, as well as the importance of developing local capital markets to facilitate long-term investments.

The distribution of climate finance in developing countries is extremely uneven. While a few large economies, such as Brazil, India, China, and South Africa, receive a sizable portion of this funding, the least developed countries (LDCs) continue to receive a much smaller share (Digitemie et al., 2024; Labatt & White, 2011). This disparity is concerning because LDCs frequently face more severe economic challenges, making it more difficult for them to secure necessary climate finance.

Despite the fact that developed countries have relatively better economic conditions, climate finance growth has been modest (Digitemie & Ekemezie, 2024). This is cause for concern because it indicates that even in countries with more financial resources, there is a reluctance to invest in climate action. This reluctance could be attributed to a variety of factors, including short-term economic priorities, political uncertainties, or a lack of understanding of the long-term costs of inaction on climate.

2.4 THE LDF

The establishment of the LDF was a watershed moment for Africa in the fight against climate change. It was a hard-fought victory that recognized the

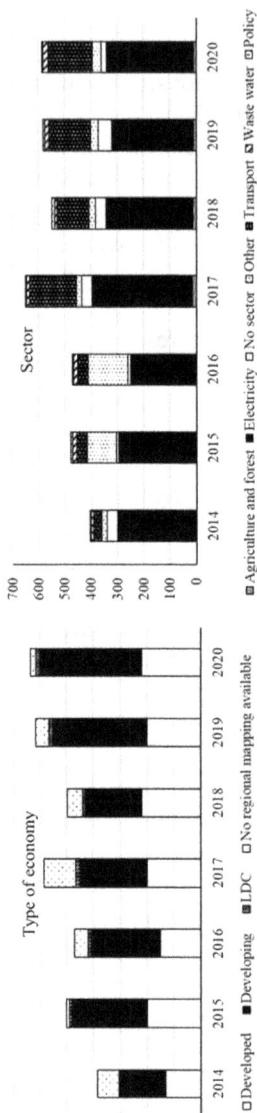

FIGURE 2.3 Available estimates of global climate finance between 2014 and 2020. (Source: IPCC report chapter 15.)

continent's extreme vulnerability to climate disasters. Africa bears the brunt of climate change, despite contributing the least to global emissions (Chirisa et al., 2021; Kissinger et al., 2019). Droughts, floods, and rising sea levels are causing havoc in African communities, displacing people, destroying livelihoods, and crippling economies.

The LDF provides a ray of hope (Salimi Turkamani, 2024). It represents a global commitment to assisting African nations dealing with the devastating effects of climate change that they did not cause (Bendell, 2018). This fund can provide critical resources for disaster recovery, displaced population resettlement, and climate-resilient infrastructure investment (Salimi Turkamani, 2024; Seymour & Busch, 2016). However, there is cause for cautious optimism (Seymour & Busch, 2016). The amount of money pledged thus far falls far short of the estimated requirements. Experts estimate that hundreds of billions of dollars are needed each year to address the losses and damages that African nations are already suffering, with the figure increasing as the climate crisis worsens (Salimi Turkamani, 2024). The LDF current funding gap threatens to undermine its very purpose (Bendell, 2018; Salimi Turkamani, 2024). There is an urgent need for developed countries, the largest historical polluters, to increase their financial contributions (Gahlawat & Lakra, 2020; Quigley, 2006). They must keep their promises and ensure that the fund has adequate resources. Only then will the LDF be able to fulfill its role as a lifeline for Africa during the climate crisis (Bendell, 2018; Salimi Turkamani, 2024).

There is a progress in the fight against climate change, but it also serves as a reminder of the enormous challenge that still exists (Gahlawat & Lakra, 2020; Seymour & Busch, 2016). At a recent climate conference, an unprecedented $85 billion was pledged for global climate action (Betts et al., 2018; Quigley, 2006; Salimi Turkamani, 2024). This substantial sum will be used to support a variety of initiatives aimed at lowering greenhouse gas emissions and mitigating the effects of climate change. However, a closer look reveals a disparity in the funding. While $85 billion is a large sum, only $792 million was specifically allocated to the LDF (Quigley, 2006; Salimi Turkamani, 2024). This fund is critical for assisting countries that are already facing the devastating effects of climate change, such as rising sea levels, extreme weather events, and agricultural disruptions (Gahlawat & Lakra, 2020; Treidel et al., 2011). These countries, often developing nations with limited resources, have made the least contribution to climate change but bear the brunt of its consequences.

The significant disparity between overall funding and the amount allocated to the LDF emphasizes the need for a more equitable approach (Bendell, 2018; Quigley, 2006). While mitigation efforts are critical to preventing future damage, helping those who are already struggling is equally important (Salimi Turkamani, 2024). Moving forward, it is critical that future funding allocations better reflect the importance of assisting vulnerable nations in dealing with the realities of climate change.

2.4.1 The Power of Money and Measurement in Combating Climate Change

Combating climate change necessitates a two-pronged approach: lowering emissions and shifting to cleaner energy sources (Boumis et al., 2023; Chirisa et al., 2021; Kissinger et al., 2019). Secured climate finance is a critical weapon in this fight. By providing financial resources, developed countries can help developing countries implement projects that reduce their reliance on fossil fuels (Fakir, 2023; Gordon, 2023). This could include providing funding for the construction of solar or wind farms, as well as promoting the development of energy-efficient technologies. These investments directly contribute to a reduction in greenhouse gas emissions, bringing us closer to our climate goals. However, simply throwing money at the problem is insufficient. Effective monitoring of Nationally Determined Contributions (NDCs) is equally important (Sachs et al., 2022). The NDCs are each country's pledge to reduce emissions and adapt to climate change (Fakir, 2023; Sachs et al., 2022). Tracking progress toward these goals allows us to identify areas where more action is required. This targeted approach ensures that climate finance is used effectively and that countries remain on track to achieve a clean energy transition. Consider it like steering a ship: secure climate finance provides the fuel, while effective NDC monitoring serves as the compass, guiding us toward a sustainable future.

2.4.2 Value Chains for Sustainable Forest Management: The Power of Directives and Guidelines

One of the pillars of a country's sustainable development strategy is to create an environment that promotes responsible resource use (Chirisa et al., 2021). In the forestry sector, this means developing a framework that balances resource development and long-term protection. Well-defined legal instruments are critical in this context.

Creating clear directives and guidelines for public-private partnerships (PPPs), Forest Carbon Financing Mechanisms, Forest Concessions, and Private Sector Engagement in Carbon Financing is an effective tool for transforming the forestry sector (Odhong' et al., 2019; VI, 2014). These documents can help to create a transparent and efficient framework for collaboration between the public and private sectors.

- **Empowering Partnerships and Investment:**
  Clear guidelines for public-private partnerships (PPPs) can spur much-needed investment in sustainable forestry practices (Odhong' et al., 2019). This enables private companies to collaborate with the government on forest management initiatives such as reforestation, tree improvement programs, and responsible harvesting techniques.
- **Unlocking the potential of carbon markets:**
  Developing effective Forest Carbon Financing Mechanisms enables the forestry sector to participate in the growing carbon market (Timilsina, 2021). These mechanisms encourage sustainable forest management practices by rewarding individuals and organizations that store carbon in healthy forests (Table 2.2).
- **Streamlining forest concessions:**
  Clear Forest Concession Guidelines establish a framework for allocating forest land to private entities for sustainable management (Table 2.2). This can encourage responsible forestry practices that result in increased forest cover and economic development (Table 2.2.).
- **Engaging the private sector in carbon finance:**
  The government can make it easier for the private sector to participate in carbon trading schemes by establishing clear guidelines for private sector engagement. This enables private companies to invest in forests, earn carbon credits, and contribute to national climate change mitigation efforts (Yoro & Daramola, 2020).

To summarize, developing these directives and guidelines provides a strategic opportunity for the forestry sector (Timilsina, 2021). These documents, which encourage collaboration, transparency, and responsible investment, can pave the way for a more sustainable future for our forests (Chirisa et al., 2021).

2.5 CLIMATE ACTION IN AFRICA

A significant step forward was recently taken to address a major issue related to climate change in Africa (Baskaran et al., 2023; Michaelowa et al., 2021). World leaders agreed to increase funding for clean cooking solutions across the continent (Yoro & Daramola, 2020). This commitment is critical for improving

TABLE 2.2   The value chains for sustainable forest management

VALUE CHAIN COMPONENT	DIRECTIVES/GUIDELINES ROLE	IMPACTS ON SUSTAINABILITY	STAKEHOLDERS INVOLVED	KEY BENEFITS
Public-Private Partnerships (PPPs)	• Setting clear terms for collaboration. • Defining responsibilities and risk-sharing. • Establishing monitoring and evaluation frameworks. • Promoting equitable benefit-sharing.	• Increased investment in reforestation, • Tree improvement, and responsible harvesting. • Enhanced implementation of sustainable forest management plans. • Technology and knowledge transfer.	• Government agencies • Private sector companies (logging, reforestation, and conservation). • Local communities. • NGOs. • Research institutions.	• Mobilized capital for sustainable initiatives. • Improved efficiency and effectiveness of forest management. • Creation of green jobs. • Stronger relationships and trust.
Forest Carbon Financing Mechanisms	• Defining criteria for carbon credit generation. • Establishing monitoring, reporting, and verification (MRV) systems. • Setting standards for carbon project development. • Facilitating access to carbon markets. • Ensuring transparency and integrity.	• Increased carbon sequestration through forest conservation and reforestation. • Financial incentives for sustainable forest management. • Reduction of deforestation and degradation. • Enhanced biodiversity conservation. • Contribution to national climate goals.	• Forest landowners and others. • Carbon project developers. • Government regulators. • Carbon credit buyers. • Verification agencies. • NGOs focusing on climate and conservation.	• Improved forest health & resilience. • Revenue generation from carbon markets. • Adoption of sustainable practices. • GHG emission reductions. • Community and PPPs financial gains.

Forest Concessions	• Establishing environmental impact assessments. • Setting sustainable harvesting. • Defining land-use planning requirements. • Ensuring compliance with labor and social standards. • Defining conditions for concessions.	• Reduced environmental impact from logging operations. • Improved forest management practices. • Increased forest cover over time. • Long-term management strategy development. • Protection of ecologically sensitive areas.	• Government forestry agencies. • Private sector logging. • Local communities living in and around concession areas. • Environmental NGOs. • Indigenous groups.	• Reduced risk of illegal logging and deforestation. • Improved forest governance. • Stable supply of timber and non-timber forest products. • Long-term planning certainty.
Private Sector Engagement in Carbon Financing	• Providing guidelines on how to participate in carbon trading schemes. • Establishing clear procedures for private sector investment in carbon projects. • Reducing regulatory barriers. • Creating incentives for private sector participation. • Standardizing contracts.	• Increased private sector investment in forest conservation and restoration. • Greater innovation in carbon project development. • Enhanced capacity for national carbon market participation. • Funding for conservation projects. • Contribution to corporate social responsibility (CSR) goals.	• Private sector companies (energy, finance, and manufacturing). • Carbon project developers and aggregators. • Government regulators. • Carbon market exchanges. • Consulting firms.	• Access to private sector financial resources. • Increased scale and reach of carbon reduction projects. • Financial incentives for investment in nature-based solutions. • Enhanced national mitigation efforts.

the lives of millions of Africans who are currently dependent on unhealthy and unsustainable cooking methods like open fires fueled by firewood.

The lack of access to safe cooking methods has far-reaching consequences. Smoke inhalation from traditional cooking fires is a major cause of respiratory illness, particularly among women and children who spend a lot of time near cooking areas (Chirisa et al., 2021; Tamasiga et al., 2023). Furthermore, the constant need for firewood contributes to deforestation, harms the environment, and increases the workloads of women, who are frequently responsible for fuel collection (Gordon, 2023; Michaelowa et al., 2021).

The Africa Climate Summit, held in Nairobi, Kenya, on September 2023, was more than just a forum for discussing the challenges Africa faces due to climate change. It grew into a powerful platform for demanding action (Timilsina, 2021). The urgent need for climate finance was a common theme throughout the summit. African nations, which are particularly vulnerable to the effects of climate change despite contributing minimally to global emissions, have requested increased financial assistance from developed countries (Yoro & Daramola, 2020). The summit emphasized the need for a complete overhaul of the global financial system, rather than simply more money.

The current architecture, it was argued, fails to address the specific needs of developing countries dealing with climate change (Adenle et al., 2017; Michaelowa et al., 2021; Teweldebrihan & Dinka, 2024). The call was for a radical transformation, a new system that prioritizes fairness and provides Africa with the resources it urgently requires to adapt to a changing climate and build a viable future (Chirisa et al., 2021; Gordon, 2023).

Two significant initiatives emerged from the recent COP conference that have the potential to transform Africa's energy landscape and achieve climate goals (Sachs et al., 2022). The first is the introduction of the Global Renewables and Energy Efficiency Pledge. This ambitious commitment aims to triple the world's renewable energy capacity by 2030. This translates into a massive shift toward cleaner sources of energy such as solar and wind. For Africa, this represents an enormous opportunity to "leapfrog" traditional fossil fuel dependence and establish a modern energy infrastructure based on renewables.

Second, the Global Methane Pledge, which gained momentum at COP, aims to reduce global methane emissions by 30% by 2030. Methane, a potent greenhouse gas, has a shorter lifespan than carbon dioxide but packs a much bigger warming punch in the short term (Chirisa et al., 2021; Mugambiwa & Kwakwa, 2022). A significant number of African countries have already pledged to join this initiative. Africa can make a significant contribution to reducing global warming and meeting its own climate targets by addressing methane emissions, particularly from sectors such as oil and gas, and waste management (Gordon, 2023; Mugambiwa & Kwakwa, 2022).

In essence, these two pledges provide Africa with a powerful one-two punch to address both its energy and environmental challenges (Chirisa et al., 2021; Savvidou et al., 2021; Tamasiga et al., 2023). By embracing renewable energy sources and reducing methane emissions, the continent can pave the way for a more sustainable and environmentally friendly future (Mungai et al., 2022).

2.6 CLIMATE FINANCE MECHANISMS IN AFRICA AND DEVELOPING COUNTRIES

The climate financing architecture in Africa and developing countries is distinguished by a continuous gap between declared commitments and actual execution (Barnard et al., 2014). While the UNFCCC requires industrialized countries to contribute to the incremental costs of climate mitigation in emerging economies, a lack of open and effective procedures has slowed progress. The significant financial flows projected from global climate negotiations, particularly to sub-Saharan Africa, raise concerns about potential negative consequences, such as volatility, Dutch disease, and rent-seeking (Harvie, 2019). These dangers, arising from both non-market direct transfers and market-based systems such as international emissions trading (IET), highlight the need for strong institutional structures to avert a "climate finance curse" (Stender et al., 2020). According to studies, IET, particularly under per capita allocation systems, has the potential to create much bigger financial transfers than non-market alternatives, but it also has higher volatility risks due to variable permit costs (Digitemie & Ekemezie, 2024).

Climate financing delivery techniques are roughly divided into two categories: non-market and market-based (Khan et al., 2020). Non-market techniques use direct transfers to cover incremental or total mitigation costs, whereas market-based approaches, like IET, rely on permit allocation procedures. The possibility of significant financial inflows, particularly through IET, needs careful study of institutional safeguards to avoid unwanted outcomes. Carbon market pricing corridors, sovereign wealth funds for long-term volatility management, and conditionality rules to combat corruption are all critical measures (Härtel & Korpås, 2021). Limiting windfall profits from climate funding, whether through changes to allocation methods or stringent fund utilization oversight, is critical to ensuring that these transfers promote sustainable development and resilience. Addressing these issues is crucial to the efficient and equitable distribution of climate finances, allowing developing countries to effectively combat climate change.

2.6.1 Climate Finance Flow and Accountability

The flow of climate funding to Africa and other developing countries, as well as the accountability that comes with it, poses a serious barrier to global efforts to combat climate change (Chirambo, 2014; Arezki, 2021). While the United Nations Framework Convention on Climate Change (UNFCCC) requires international financial aid to help vulnerable regions adapt to climate impacts, the reality on the ground frequently falls short of expectations (Kinley et al., 2021). The efficacy of this financial aid is dependent on both the amount of monies raised and the strategic distribution of those resources (Songwe *et al.*et al., 2022). However, there is a significant vacuum in the absence of a thorough quantitative analysis that precisely maps adaptation-related financing flows to African countries. This lack of openness impedes proper monitoring and assessment, making it difficult to determine the true impact of these financial pledges. A rigorous review of development funding, specifically adaptation finance, from bilateral and multilateral funders to Africa between 2014 and 2018 uncovers some troubling trends.

The different study shows that the actual amounts of cash given are much lower than the large investments required for adaptation in a continent that is very vulnerable to climate change and has poor adaptation capacity. Notably, financing for mitigation initiatives (US$30.6 billion) almost exceeded that budgeted for adaptation (US$16.5 billion). This difference demonstrates a potential imbalance in addressing African nations' pressing needs, where adaptability is frequently paramount. Furthermore, the distribution of these money differs significantly between African countries, indicating potential discrepancies in resource allocation. The fact that the bulk of adaptation-related financing was provided in the form of loans (57%) rather than grants (42%), raises concerns about increased debt loads for already vulnerable countries. Furthermore, the concentration of half of the adaptation money in just two sectors, agriculture and water supply and sanitation, implies a potential lack of diversification in dealing with the varied effects of climate change (Buchner *et al.*et al., 2014). A particularly concerning data is the disbursement ratio for adaptation during this period, which was only 46%, much lower than the 96% disbursement percentage for total African development money. This disparity highlights the difficulties in turning financial commitments into concrete activities on the ground. The study leverages data from two OECD databases that painstakingly track bilateral and multilateral funder support to African governments, mostly through development finance organizations. However, the ongoing ambiguity around the notion of "new and additional" resources presents a substantial analytical challenge. This word, which is critical for African states under the UNFCCC framework, lacks a precise, widely acknowledged meaning, making it impossible to definitively determine whether reported adaptation financing genuinely enhances existing development aid. As a result, the study displays the total reported adaptation funds without proving their additionality, recognizing the intrinsic restrictions imposed by the definitional gap. This lack of clarity affects accountability and transparency in climate finance, underlining the critical need for a defined definition and rigorous monitoring methods to guarantee that financial commitments result in meaningful, long-term benefits for African communities.

2.6.2 Climate Fund Disbursement Rates

The flow of climate funding to Africa and other developing countries, as well as the accountability that comes with it, poses a serious barrier to global efforts to combat climate change (Arezki, 2021). While the United Nations Framework Convention on Climate Change (UNFCCC) requires international financial aid to help these vulnerable regions adapt to climate change, the reality on the ground frequently falls short of expectations (Kinley *et al.*et al., 2021). The success of this financial aid is determined by both the volume of funds mobilized and the strategic distribution of those resources. However, a critical gap persists in the absence of a thorough quantitative analysis that precisely maps adaptation-related financing flows to African countries. This lack of openness stifles effective monitoring and evaluation, making it difficult to measure the true impact of these financial investments. A rigorous review of development funding, specifically adaptation finance, from bilateral and multilateral funders to Africa between 2014 and 2018 uncovers some troubling trends (Buhr *et al.*et al., 2018; Mugambiwa & Kwakwa, 2022).

Different studies showed that the actual amounts of cash given are much lower than the large investments required for adaptation in a continent that is very vulnerable to climate change and has poor adaptation capacity (Buhr et al., 2018; Hammill et al., 2008; Kinley et al., 2021; Savvidou et al., 2021). Notably, financing for mitigation initiatives (US$30.6 billion) almost exceeded that budgeted for adaptation (US$16.5 billion). This difference demonstrates a potential imbalance in addressing African nations' pressing needs, where adaptability is frequently paramount (Mungai et al., 2021). Furthermore, the distribution of this money differs significantly between African countries, indicating potential discrepancies in resource allocation (Table 2.3). The fact that the bulk of adaptation-related financing was provided in the form of loans (57%) rather than grants (42%), raises concerns about increased debt loads for already vulnerable countries (Savvidou *et al.*et al., 2021). Furthermore, the concentration of half of the adaptation money in just two sectors, agriculture and water supply and sanitation, implies a potential lack of diversification in dealing with the varied effects of climate change (Table 2.3). A particularly

TABLE 2.3   The climate fund disbursement analysis (Africa, 2014–2018)

METRIC	FINDING	IMPLICATION
Total Adaptation Funding	• US$16.5 billion	• Significantly lower than required adaptation investment.
Total Mitigation Funding	• US$30.6 billion	• Imbalance favoring mitigation over adaptation, despite Africa's immediate adaptation needs.
Funding Distribution	• Uneven distribution across African countries	• Potential inequities in resource allocation; need for improved targeting based on vulnerability.
Loan vs. Grant Ratio (Adaptation)	• 57% loans, 42% grants	• Increased debt burden for vulnerable countries; preference for grants is needed.
Sectoral Concentration (Adaptation)	• 50% in agriculture and water supply/sanitation	• Lack of diversification; neglect of other climate-vulnerable sectors.
Adaptation Disbursement Rate	• 46%	• Significant delays in translating financial commitments into on-the-ground action; highlights implementation challenges.
Overall African Development Disbursement Rate	• 96%	• Shows that there is a specific problem with climate adaptation funding dispersal.

concerning data is the disbursement ratio for adaptation during this period, which was only 46%, much lower than the 96% disbursement percentage for total African development money (Mugambiwa & Kwakwa, 2022; Savvidou et al., 2021). This disparity highlights the difficulties in turning financial commitments into concrete activities on the ground.

2.6.3 Completion Rate of Climate Finance Projects

The 46% completion rate, or disbursement ratio, for adaptation climate money in Africa between 2014 and 2018 demonstrates a serious failure to translate cash promises into on-the-ground action. This ratio, which contrasts sharply with the 96% disbursement rate for overall African development funding, highlights a substantial bottleneck unique to climate adaptation. This gap is more than just a bureaucratic snag; it indicates that less than half of the committed monies designed to strengthen Africa's resilience to climate change really reached their intended recipients. As a result, the analysis shows the total funds reported by funders to the OECD as assisting adaptation, recognizing that it is impossible to assess whether these amounts are indeed supplementary to pre-existing development finance commitments. This ambiguity complicates efforts to ensure accountability and transparency in climate finance flows, highlighting the need for clearer definitions and improved monitoring mechanisms to ensure that financial commitments result in meaningful and long-term change for African communities.

2.7 CONCLUSION

The report and study analysis reveal significant challenges in climate finance mechanisms for Africa and developing countries. Primarily, there is a disconnect between pledged financial commitments and their actual implementation, hampered by a lack of transparency and effective distribution methods. Large-scale climate finance inflows, while potentially beneficial, pose risks like economic volatility, Dutch disease, and corruption, demanding robust institutional safeguards. Furthermore, the allocation of funds reveals a concerning imbalance, with mitigation efforts receiving more funding than adaptation, despite adaptation being crucial for vulnerable regions. A significant portion of adaptation finance is disbursed as loans, increasing debt burdens, and the disbursement rate itself is alarmingly low, indicating difficulties in translating financial pledges into tangible action. Crucially, the absence of a clear definition for "new and additional" resources within the UNFCCC framework creates ambiguity, hindering accountability and transparency. In conclusion, the efficacy of climate finance in Africa and developing countries is severely compromised by inconsistent implementation, potential economic risks, funding imbalances, debt-inducing loans, low disbursement rates, and a lack of clear definitions, necessitating stronger institutional frameworks, improved transparency, and a focus on adaptation to ensure effective and sustainable climate action.

REFERENCES

Adenle, A. A., Manning, D. T., & Arbiol, J. (2017). Mitigating climate change in Africa: Barriers to financing low-carbon development. *World Development, 100*, 123–132.

Arezki, R. (2021). Climate finance for Africa requires overcoming bottlenecks in domestic capacity. *Nature Climate Change*, *11*(11), 888–888.

Barnard, S., Watson, C., Greenhill, R., Caravani, A., Trujillo, N. C., Hedger, M., & Whitley, S. (2014). *Climate finance: Is it making a difference?* Overseas Development Institute.

Baskaran, G., Ekeruche, A., Heitzig, C., Ordu, A. U., & Senbet, L. W. (2023). Financing climate-resilient infrastructure in Africa.

Belianska, A., Bohme, N., Cai, K., Diallo, Y., Jain, S., Melina, M. G., ... Zerbo, S. (2022). *Climate change and select financial instruments: An overview of opportunities and challenges for Sub-Saharan Africa*. International Monetary Fund.

Bendell, J. (2018). *Deep adaptation: A map for navigating climate tragedy*. University of Cumbria.

Betts, R. A., Alfieri, L., Bradshaw, C., Caesar, J., Feyen, L., Friedlingstein, P., ... Morfopoulos, C. (2018). Changes in climate extremes, fresh water availability and vulnerability to food insecurity projected at 1.5 C and 2 C global warming with a higher-resolution global climate model. *Philosophical Transactions of the Royal Society A: Mathematical, Physical and Engineering Sciences*, *376*(2119), 20160452.

Boumis, G., Moftakhari, H. R., & Moradkhani, H. (2023). Coevolution of extreme sea levels and sea-level rise under global warming. *Earth's Future*, *11*(7), e2023EF003649.

Buchner, B., Stadelmann, M., Wilkinson, J., Mazza, F., Rosenberg, A., & Abramskiehn, D. (2014). Global landscape of climate finance 2019. *Climate Policy Initiative*, *32*(1), 1–38.

Buhr, B., Volz, U., Donovan, C., Kling, G., Lo, Y. C., Murinde, V., & Pullin, N. (2018). *Climate change and the cost of capital in developing countries*. UN Environment, Imperial College London and SOAS University of London.

Chirambo, D. (2014). The climate finance and energy investment dilemma in Africa: Lacking amidst plenty. *Journal of Developing Societies*, *30*(4), 415–440.

Chirambo, D. (2016). Integrating microfinance, climate finance and climate change adaptation: A sub-Saharan Africa perspective. In *Climate change adaptation, resilience and hazards* (pp. 195–207). Springer Nature.

Chirambo, D. (2017). Enhancing climate change resilience through microfinance: Redefining the climate finance paradigm to promote inclusive growth in Africa. *Journal of Developing Societies*, *33*(1), 150–173.

Chirisa, I., Gumbo, T., Gundu-Jakarasi, V. N., Zhakata, W., Karakadzai, T., Dipura, R., & Moyo, T. (2021). Interrogating climate adaptation financing in Zimbabwe: Proposed direction. *Sustainability*, *13*(12), 6517.

Digitemie, W. N., & Ekemezie, I. O. (2024). Assessing the role of climate finance in supporting developing nations: A comprehensive review. *Finance & Accounting Research Journal*, *6*(3), 408–420.

Doku, I., Ncwadi, R., & Phiri, A. (2021). Determinants of climate finance: Analysis of recipient characteristics in sub-Sahara Africa. *Cogent Economics & Finance*, *9*(1), 1964212.

Fakir, S. (2023). South Africa's just energy transition partnership: A novel approach transforming the international landscape on delivering NDC financial goals at scale. *South African Journal of International Affairs*, *30*(2), 297–312.

Gahlawat, I. N., & Lakra, P. (2020). Global climate change and its effects. *Integrated Journal of Social Sciences, 7*(1), 14–23.

Gordon, N. J. (2023). Climate finance: An overview. *Environment: Science and Policy for Sustainable Development, 65*(4), 18–26.

Hammill, A., Matthew, R., & McCarter, E. (2008). Microfinance and climate change adaptation. *IDS Bulletin, 39*(4), 113–122.

Härtel, P., & Korpås, M. (2021). Demystifying market clearing and price setting effects in low-carbon energy systems. *Energy Economics, 93*, 105051.

Harvie, C. (2019). The Dutch disease and economic diversification: Should the approach by developing countries be different? *Trade Logistics in Landlocked and Resource Cursed Asian Countries, 13*(1), 9–45.

Hidayati, I. (2021). Migration as a coping strategy of Indonesian farmers in the face of climate change. In Paper presented at the IOP Conference Series: Earth and Environmental Science.

IPCC, C. C. J. C. U. P., Cambridge, United Kingdom, & New York, N., USA. (2007). *The physical science basis. Contribution of working group I to the fourth assessment report of the intergovernmental panel on climate change* (p. 996). Cambridge University Press.

Khan, M., Robinson, S.-A., Weikmans, R., Ciplet, D., & Roberts, J. T. (2020). Twenty-five years of adaptation finance through a climate justice lens. *Climatic Change, 161*(2), 251–269.

Kinley, R., Cutajar, M. Z., de Boer, Y., & Figueres, C. (2021). Beyond good intentions, to urgent action: Former UNFCCC leaders take stock of thirty years of international climate change negotiations. *Climate Policy, 21*(5), 593–603.

Kissinger, G., Gupta, A., Mulder, I., & Unterstell, N. (2019). Climate financing needs in the land sector under the Paris Agreement: An assessment of developing country perspectives. *Land Use Policy, 83*, 256–269.

Labatt, S., & White, R. R. (2011). *Carbon finance: The financial implications of climate change*. John Wiley & Sons.

Mekonnen, A. (2014). Economic costs of climate change and climate finance with a focus on Africa. *Journal of African Economies, 23*(suppl_2), ii50–ii82.

Michaelowa, A., Hoch, S., Weber, A.-K., Kassaye, R., & Hailu, T. (2021). Mobilising private climate finance for sustainable energy access and climate change mitigation in sub-Saharan Africa. *Climate Policy, 21*(1), 47–62.

Mugambiwa, S., & Kwakwa, M. (2022). Multilateral climate change financing in the developing world: Challenges and opportunities for Africa. *International Journal of Research in Business and Social Science (2147-4478), 11*(9), 306–312.

Mungai, E. M., Ndiritu, S. W., & Da Silva, I. (2021). Unlocking climate finance potential for climate adaptation: Case of climate smart agricultural financing in Sub Saharan Africa. In *African handbook of climate change adaptation* (pp. 2063–2083). Springer.

Mungai, E. M., Ndiritu, S. W., & Da Silva, I. (2022). Unlocking climate finance potential and policy barriers – A case of renewable energy and energy efficiency in sub-Saharan Africa. *Resources, Environment and Sustainability, 7*, 100043.

Nakhooda, S., Caravani, A., Bird, N., Schalatek, L., & America, H. (2011). *Climate finance in sub-Saharan Africa*. Climate Finance Policy Briefs, Heinrich Böll Stiftung North America, Washington, DC, USA and Overseas Development Institute (ODI).

Odhong', C., Wilkes, A., van Dijk, S., Vorlaufer, M., Ndonga, S., Sing'ora, B., & Kenyanito, L. (2019). Financing large-scale mitigation by smallholder farmers: What roles for public climate finance? *Frontiers in Sustainable Food Systems, 3*, 3.

Quigley, W. P. (2006). Thirteen ways of looking at Katrina: Human and civil rights left behind again. *Tulane Law Review, 81*, 955.

Rossitto, N. (2021). Green bonds: An alternative source of financing in the era of climate change.

Sachs, J. D., Toledano, P., Dietrich Brauch, M., Mebratu-Tsegaye, T., Uwaifo, E., & Sherrill, B. M. (2022). Roadmap to zero-carbon electrification of Africa by 2050: The green energy transition and the role of the natural resource sector (minerals, fossil fuels, and land).

Salimi Turkamani, H. (2024). The loss and damage fund: A solution to interpretive conflicts of responsibility for climate change? *Netherlands International Law Review, 71*(3), 1–26.

Savvidou, G., Atteridge, A., Omari-Motsumi, K., & Trisos, C. H. (2021). Quantifying international public finance for climate change adaptation in Africa. *Climate Policy, 21*(8), 1020–1036.

Seymour, F., & Busch, J. (2016). *Why forests? Why now?: The science, economics, and politics of tropical forests and climate change*. Brookings Institution Press.

Shirai, S. (2023). *Global climate challenges, innovative finance, and green central banking*. Asian Development Bank.

Songwe, V., Stern, N., & Bhattacharya, A. (2022). *Finance for climate action: Scaling up investment for climate and development*. Grantham Research Institute on Climate Change and the Environment, London School of Economics and Political Science.

Stender, F., Moslener, U., & Pauw, W. P. (2020). More than money: Does climate finance support capacity building? *Applied Economics Letters, 27*(15), 1247–1251.

Tamasiga, P., Molala, M., Bakwena, M., Nkoutchou, H., & Onyeaka, H. (2023). Is Africa left behind in the global climate finance architecture: Redefining climate vulnerability and revamping the climate finance landscape – A comprehensive review. *Sustainability, 15*(17), 13036.

Teweldebrihan, M., & Dinka, M. (2024). The impact of climate change on the development of water resources. *Global Journal of Environmental Science and Management*. 10(3), 1359–1370.

Timilsina, G. R. (2021). Financing climate change adaptation: International initiatives. *Sustainability, 13*(12), 6515.

Treidel, H., Martin-Bordes, J. L., & Gurdak, J. J. (2011). *Climate change effects on groundwater resources: A global synthesis of findings and recommendations*. CRC Press.

VI, P. O. H. K. M. (2014). *Climate financing: Implications for Africa's transformation*. United Nations Economic Commission for Africa (UNECA).

Yoro, K. O., & Daramola, M. O. (2020). CO_2 emission sources, greenhouse gases, and the global warming effect. In *Advances in carbon capture* (pp. 3–28). Elsevier.

Status Quo of African Countries on Climate Financing

3

3.1 INTRODUCTION

Climate change poses a substantial and diverse threat to human health and well-being worldwide, with African healthcare systems experiencing especially catastrophic implications (Butler, 2018; Garrett, 2013; Møller & Roberts, 2021). The UAE Framework for Global Climate Resilience (UAE FGCR) is a critical initiative that establishes a system for tracking progress on adaptation efforts and directing us to the most effective solutions, in which the emphasis is mainly on three major areas (Angerer et al.; Njuguna et al., 2024; Winkler et al., 2024a). To properly handle this situation, a comprehensive strategy is required. This approach focuses on three major areas: improving monitoring and measuring, emphasizing people-centered adaptation, and investing in adaptation financing. The UAE FGCR prioritizes the creation of quantitative metrics for assessing countries' adaptation to climate change (Njuguna et al., 2024; Savvidou et al., 2021). This method guarantees that adaptation efforts are directed toward the most pressing requirements, resulting in a more coordinated global response.

DOI:10.1201/9781003644378-3

Strengthening Monitoring and Measurement:
- The UAE FGCR emphasizes the development of quantifiable indicators to assess countries' adaptation to climate change (Angerer et al.; Njuguna et al., 2024; Scullion et al., 2019). This will ensure that adaptation efforts are focused on the most pressing issues, promoting a coordinated global response (Figure 3.1).

People-Centered Adaptation:
- Addressing the effects of climate change on health necessitates a two-pronged strategy (Angerer et al.; Bogale & Erena, 2022; Oladosu & Chanimbe, 2024). First, empower communities by involving them in the development of solutions that address their specific needs. Second, address underlying health disparities by providing clean water, sanitation, and nutritious food (Mosadeghrad et al., 2023; Seltenrich, 2018).

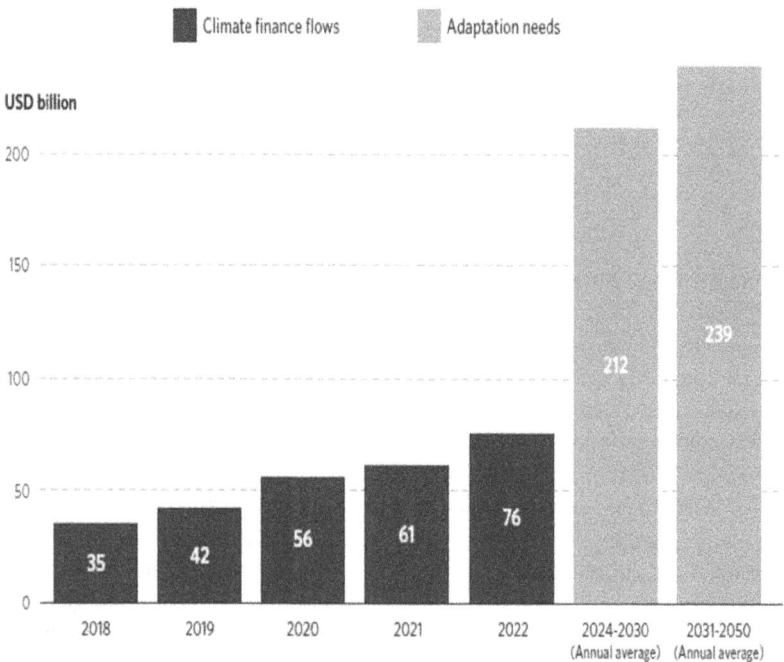

FIGURE 3.1   The patterns in global climate adaptation finance flows versus climate finance adaptation requirements. (Source CPI, 2024.)

Investment in Adaptation Financing:

- Increased financial resources, specifically for the African health sector, are critical (Butler, 2018; Garrett, 2013; Oladosu & Chanimbe, 2024). This funding can be used to improve early warning systems, build resilient healthcare infrastructure, and train healthcare workers to deal with climate-related health risks (Figure 3.1).

Furthermore, tackling the health consequences of climate change necessitates a dual strategy that empowers communities while addressing underlying health inequities. Community empowerment entails actively engaging individuals and communities in producing solutions that are tailored to their specific needs. Addressing health disparities includes providing access to clean water, sanitation, and nutritious food, all of which are necessary for optimal health.

Finally, improved financial resources, particularly in the African health sector, are critical. This cash can be used to develop early warning systems, construct robust hospital infrastructure, and train healthcare professionals to address climate-related health risks. An integrated approach that includes these methods is critical for developing a just and equitable response to climate change challenges and guaranteeing a healthier future for everyone. Furthermore, nations such as Ethiopia are displaying ambitious climate action through their Nationally Determined Contributions (NDCs), which aim for significant emissions reductions while also investing in mitigation and adaptation efforts. These efforts emphasize the significance of integrating international assistance with national actions to effectively address climate change and protect populations' health, particularly in vulnerable countries such as Africa and developing countries. Therefore, an integrated approach, which includes the abovementioned implementation strategies, might create a more just and equitable response to climate change threats while also ensuring a healthier future for all.

3.2 RATIFICATION OF GLOBAL ENVIRONMENTAL AGREEMENTS

The UAE Framework for Global Climate Resilience (UAE FGCR) is laying the groundwork for a critical aspect of climate change mitigation: tracking progress on adaptation initiatives (Fleishel et al., 2021; Winkler et al., 2024a). This ongoing work is necessary for a variety of reasons (Gordon, 2023; Seltenrich,

2018). First, the UAE FGCR intends to lay a solid foundation for the development of quantifiable indicators and metrics. These are essentially the tools we'll use to assess how well countries adapt to the effects of climate change (Briera & Lefèvre, 2024; Fleishel et al., 2021). Without clear metrics, it is difficult to assess progress and identify areas where we should increase our efforts. The UAE FGCR's emphasis on strong and credible metrics will ensure that these tools are accurate and useful (Angerer et al.; Winkler et al., 2024b). Second, the framework emphasizes focusing adaptation efforts on the most pressing issues.

3.3 FRAMEWORK FOR CLIMATE-RESILIENT HEALTH: PRIORITIZING PEOPLE AND EQUITY

Climate change is already having a significant impact on global health and well-being, and the risks are expected to worsen (Chirisa et al., 2021; Garrett, 2013; Gordon, 2023; Seltenrich, 2018). By focusing on relevant priorities, actions, and outcomes, the UAE FGCR ensures that adaptation efforts are directed to areas where they are most urgently needed (Angerer et al.; Fleishel et al., 2021; Njuguna et al., 2024). This will help to protect people's health and well-being while also increasing overall resilience in the face of climate change (Fleishel et al., 2021; Winkler et al., 2024a).

In essence, the UAE FGCR is developing a system to monitor progress and direct us toward the most effective adaptation strategies (Fleishel et al., 2021; Gordon, 2023; Winkler et al., 2024b). This will be critical to ensuring a coordinated and effective global response to climate change(Stenchikov, 2021). In the fight for climate justice, addressing climate change's impact on health necessitates a two-pronged approach (Chirisa et al., 2021; Møller & Roberts, 2021). First, we must empower communities by emphasizing people-centered solutions. This entails actively involving individuals and communities most vulnerable to climate change in developing solutions that address their unique needs and circumstances (Bogale & Erena, 2022; Mikulewicz, 2018). Second, we must address the underlying causes of health disparities, which climate change exacerbates. This includes providing access to clean water, adequate sanitation facilities, and a consistent supply of healthy and nutritious food (Mosadeghrad et al., 2023; Oladosu & Chanimbe, 2024). These basic requirements are the foundation of good health, and climate change threatens to disrupt or destroy them for many people (Fleishel et al., 2021; Mosadeghrad et al., 2023). By focusing on these two critical areas, we can develop a more just and equitable response to the threats posed by our changing climate (Briera & Lefèvre, 2024).

3.4 IMPLEMENTATION OF NATIONAL POLICIES AND STRATEGIES

The IPCC's 6th Assessment Report paints a bleak picture of Africa's healthcare systems under the growing threat of climate change (Gordon, 2023; Mugambiwa & Kwakwa, 2022; Oladosu & Chanimbe, 2024). These systems, which are already burdened by limited resources, infrastructure, and healthcare professionals, will face even greater challenges as the climate changes. There are several factors that contribute to this vulnerability. Climate change is expected to speed up the spread of climate-sensitive diseases such as malaria, dengue fever, and malnutrition (Butler, 2018; Chirisa et al., 2021). Rising temperatures and changing weather patterns provide ideal breeding conditions for mosquitoes and other disease vectors. Furthermore, extreme weather events such as floods and droughts can disrupt healthcare delivery and damage critical infrastructure, reducing access to care.

To combat these threats, increased adaptation financing, specifically directed at the African health sector, is critical (Baskaran et al., 2023; Gordon, 2023; Odhong' et al., 2019). This influx of capital can be used to improve the resilience of African health systems in a variety of ways. Early warning systems for climate-related outbreaks can be improved, allowing for more timely and targeted responses (Gordon, 2023; Mikulewicz, 2018; Omukuti, 2024). Investments can also be made to create more resilient healthcare infrastructure, such as facilities that can withstand extreme weather events (Møller & Roberts, 2021; Mikulewicz, 2018). Finally, healthcare workers can be trained to address the specific health risks associated with climate change, ensuring they have the necessary knowledge and skills to provide effective care (Butler, 2018; Mosadeghrad et al., 2023).

Investing in adaptation now can save lives, reduce long-term healthcare costs, and ensure a healthier future for African populations (Gordon, 2023; Mikulewicz, 2018; Stender et al., 2020). This is about more than just public health; it's also about economic development and overall well-being. A healthy Africa is a more resilient Africa, better prepared to face the challenges of a changing climate (Winkler et al., 2024a).

3.5 AMBITION AND PLANS ON CLIMATE FINANCING THE CASE OF ETHIOPIA

The global fight against climate change requires broad participation, and African countries, particularly Ethiopia, are making significant contributions through ambitious NDCs (Baskaran et al., 2023; Omukuti, 2024). Ethiopia's commitment is particularly impressive, with a target of a 68.8% decrease in GHG emissions by 2030, requiring a significant $316 billion investment over the next decade, with 80% going to mitigation and 13% to adaptation (Boespflug, 2024). This NDC, together with the Long-term Low Emission Development Strategy (LT-LEDS), which aims for carbon neutrality by 2050, emphasizes revolutionary reforms in the energy and forestry sectors, offering both emission reductions and economic growth (Boespflug, 2024; Winkler et al., 2024a). Ethiopia's integrated strategy includes improving agricultural methods, safeguarding and restoring forests, developing renewable energy, and increasing energy efficiency, all supported by a strong institutional framework. The reorganized Ministry of Planning and Development (MoPD) is key to coordinating these efforts and easing access to international climate funds, both of which are critical for Ethiopia's ambitious climate goals.

3.5.1 Climate Change Threatens Ethiopia's Middle-Income Goals

Studies indicate that Ethiopia's ambitious goal of becoming middle-income is inextricably linked to its agricultural sector, which serves as the foundation of the country's economy (Amare et al., 2024; Ewbank et al., 2019; Shkabatur et al., 2022). Unfortunately, the looming threat of climate change poses a serious threat to this goal (Amare et al., 2024; Ewbank et al., 2019). The negative effects of climate change on agricultural productivity, particularly in rain-fed agriculture, are threatening Ethiopia's progress (Bogale & Erena, 2022; Michaelowa et al., 2021). Thus, climate change presents a multifaceted set of challenges (Amare et al., 2024; Scott et al., 2018). Rising temperatures, which are expected to rise by an average of 0.5–2°C by 2050, will make agriculture more difficult (Michaelowa et al., 2021; Rumble & First, 2021). The limited capacity of agro-meteorological services within agricultural extension exacerbates the problem, preventing farmers from making informed decisions based on accurate weather data. El Niño and La Niña events cause highly variable rainfall patterns, creating uncertainty in agricultural planning (Amare et al., 2024; Ewbank et al., 2019). While some areas may see slightly warmer and wetter conditions, the overall trend is for more unpredictable and extreme weather events (Ewbank et al., 2019).

Climate change has a wide-ranging impact on Ethiopia's agricultural productivity (Bogale & Erena, 2022). Changes in agro-ecological zoning and agricultural practices will be required to adapt to changing climatic conditions (Angerer et al., 2024). However, these adaptations come at a high cost, with GDP potentially falling by up to 10% by 2050 (Boespflug, 2024; Michaelowa et al., 2021). If climatic conditions worsen beyond expectations, the economic impact could be far greater, exceeding 10% of GDP. Finally, climate change presents a formidable challenge to Ethiopia's aspirations for middle-income status.

The negative effects on agricultural productivity, particularly in rain-fed agriculture, threaten to stall the country's economic progress (Asfaw et al., 2019; Bogale & Erena, 2022). Addressing these issues will necessitate a comprehensive and coordinated approach that includes investments in climate-resilient agriculture, better weather forecasting, and adaptation strategies (Bendell, 2018; Bogale & Erena, 2022). Failure to act decisively could have serious consequences for Ethiopia's economic stability and the livelihoods of its citizens (Shkabatur et al., 2022).

3.5.2 Ethiopia's Ambitious Climate Action Plan

The Ethiopian National Adaptation Plan (NAP-ETH) is strategically aligned with the Climate-Resilient Green Economy (CRGE) strategy by prioritizing adaptation measures in those sectors (Conway & Schipper, 2011; Michaelowa et al., 2021; Timilsina, 2021). This synergy guarantees that existing development initiatives are strengthened by climate resilience. However, the NAP-ETH's mandate goes beyond the CRGE's. Recognizing the varied nature of climate change consequences, it will set precise targets and full implementation plans for sectors identified as particularly susceptible, even if they lie outside the CRGE's major emphasis (Table 3.1). This larger strategy aims to build a comprehensive and resilient national framework that addresses Ethiopia's multiple climate concerns.

The NAP-ETH formulation process is methodically structured, beginning with the formation of a high-level inter-ministerial steering council and the assignment of the MEFCC to lead coordination. This process includes a thorough evaluation of existing climate-related documents, such as CRGE sector plans, regional strategies, the Intended Nationally Determined Contribution (INDC), and past national statements (Table 3.2). The preparatory steps include identifying important information, policy, and strategy papers, culminating in

TABLE 3.1   The Ethiopian national adaptation plan processes

FEATURES OF NAP-ETH	DESCRIPTION OF THE STEPS AND PROCESSES
Relationship to the Climate-Resilient Green Economy (CRGE) Strategy	• Supports the Climate-Resilient Green Economy (CRGE) initiative. • Extends beyond the CRGE's scope. • Keeps a focused emphasis on adaptation. • Creates targets and implementation strategies for all vulnerable sectors, not only those identified by CRGE.
The Formulation Approach	• Incorporates climate change adaptation with current development projects. • Explores connections between climate change impacts and adaptation strategies. • Seeks synergy between climate resiliency and sustainable development. • Implements an organized method for integration.
Initiation and Mandatory Setting	• Formed an inter-ministerial steering body to provide monitoring. • The Ministry of Environment, Forest, and Climate Change (MEFCC) is responsible for coordination, leadership, and monitoring. • Established a multidisciplinary technical team for development and monitoring.
Identification of the inputs	• Existing climate change plans were reviewed (CRGE sectoral and regional plans). • Relevant documents reviewed were INDC (2015), Second National Communication (2015), EPACC (2010), and NAPA (2008). • Ensured that the NAP-ETH draws on current knowledge and activities.
The Preparation Phase	• Identified pertinent information, policy and strategy papers. • Stakeholder interaction resulted in a draft document. Conducted consultations and targeted conversations with appropriate entities. • The draft was refined to ensure that it was thorough and practical. • Conduct external consultations with all stakeholders, including sectors, regions, and key institutions and players.
Strategies for implementation	• Create guidelines for incorporating adaptation into national planning, sectoral and regional strategies, and action plans. • Identify adaptation finance mobilization techniques.

TABLE 3.2   Data sources for Ethiopia's national adaptation plan (NAP-ETH)

INFORMATION SOURCE	KEY CONTRIBUTIONS TO NAP-ETH	STATUS/NOTES
Growth and Transformation Plan II (GTP II)	• Creates a framework for incorporating climate resilience into larger development goals.	• The National five-year development plan
Ethiopia's Second National Communication to the UNFCCC	• Provides a thorough evaluation of Ethiopia's GHG inventory and includes mitigation and adaptation measures.	• The document provides data and insights.
Climate-Resilient Green Economy (CRGE) Strategy	• Identifies mitigation alternatives with adaptation benefits and presents a 20-year vision for a carbon-neutral economy.	• 20-year vision
Climate Change Impacts, Vulnerability, and Adaptation Options Report	• It provides critical data on climate change scenarios, vulnerability levels, and adaptation options.	• Prepared in collaboration with sector representatives and national professionals, utilizing IPCC GHG inventory software.
Regional and Sectoral Adaptation Plans	• It offers valuable insights into region-specific and sector-specific adaptation needs.	• Draft form
Ethiopian Programme of Adaptation to Climate Change (EPACC)	• Summarizes adaptation programs based on earlier plans, highlighting 29 prioritized climate change adaptation measures.	• The document summarizes adaptation programs.
National Adaptation Programme of Action (NAPA)	• It provided an initial foundation for adaptation planning.	• Short-term, project-based plan
Ethiopia's Intended Nationally Determined Contribution (INDC)	• It outlines major adaptation options categorized under drought, flood, and cross-sector issues.	• It contributes to the NAP-ETH's strategic direction.

the creation of a draft document through broad stakeholder engagements and targeted discussions. This methodical methodology ensures that the NAP-ETH is supported by strong evidence and reflects Ethiopian society's different needs and perspectives.

3.5.3 The Need of Building Technical Capacity for Climate Finance Mobilization

The government must prioritize the technical capabilities of key ministries, particularly the Ministry of Finance and Economic Cooperation to effectively mobilize climate finance (Bowman & Minas, 2019; Teweldebrihan & Dinka, 2024; Winkler et al., 2024b). This involves developing capacity in a variety of areas, including GCF access modalities (Bowman & Minas, 2019; Rumble & First, 2021).

Key Areas of Capacity Building Need to be Included:

Project pipeline development: It entails building a robust pipeline of climate finance projects that are consistent with national climate goals and meet the requirements of international funding mechanisms such as the Green Climate Fund.

Environmental and social safeguards: It ensures that climate-focused projects follow strict environmental and social guidelines, reducing negative impacts and promoting sustainable development.

Economic and financial analysis: It entails conducting thorough economic and financial assessments of GCF projects to determine their viability, cost-effectiveness, and potential returns.

Climate budget tagging and tracking: It refers to the implementation of systems to track climate-related expenditures within the national budget, ensuring that resources are allocated effectively and transparently.

Non-traditional climate finance: It entails identifying and obtaining alternative sources of climate finance, such as debt for climate swaps, to supplement traditional funding streams.

To Meet these Goals, the Government should:

Prepare detailed project proposals: Create comprehensive proposals for the GCF and Loss and Damage Funds, outlining project objectives, expected outcomes, and funding requirements.

Conduct economic and financial analysis: Conduct a thorough economic and financial assessment of GCF projects to determine their viability, cost-effectiveness, and potential returns.

Develop project management skills: Develop stakeholders' project management skills at the subnational and federal levels so that they can effectively implement climate projects.

Implement an IT-based platform: Create an IT-based reporting and communication platform to improve data collection, analysis, and sharing.

Implement climate budget tagging and tracking: Implement a climate budget tagging and tracking system to keep track of how climate finance is allocated and used.

Map and access non-traditional financing: Discover and investigate non-traditional climate financing options, such as public-private partnerships, forest carbon financing, and private sector participation in carbon markets.

Furthermore, by building technical capacity and implementing these strategies, the government can improve its ability to mobilize climate finance, promote sustainable development, and address climate change challenges.

3.6 CLIMATE FINANCE PROJECT IMPLEMENTATION AND TRACKING

Effective implementation and tracking of climate finance projects are critical for ensuring that allocated monies result in actual climate action (Gordon, 2023; Rumble & First, 2021). The lack of robust tracking methods makes it difficult to determine whether financial promises, such as the anticipated USD 100 billion yearly pledge to developing countries, are having a meaningful impact on emissions reduction or climate resilience(Digitemie & Ekemezie, 2024; Gordon, 2023). This weakness is exacerbated by the absence of established definitions of "climate finance," resulting in uncertainties in evaluating financial transfers and keeping parties accountable (Figure 3.2). The challenge also includes technical components, such as the requirement for precise definitions of finance flows and sources, reliable causal attribution between funding and climate effects, and well-defined boundaries for climate-related finance (Barnard et al., 2014). These data gaps and definitional inconsistencies impede the development of a transparent and dependable tracking system, which is critical for governments and the private sector to make informed decisions about future allocations and, ultimately, drive significant progress in addressing the global climate crisis (Figure 3.2).

The lack of publicly available financial statistics on NGO-led projects in Africa and other developing countries creates a substantial barrier to knowing

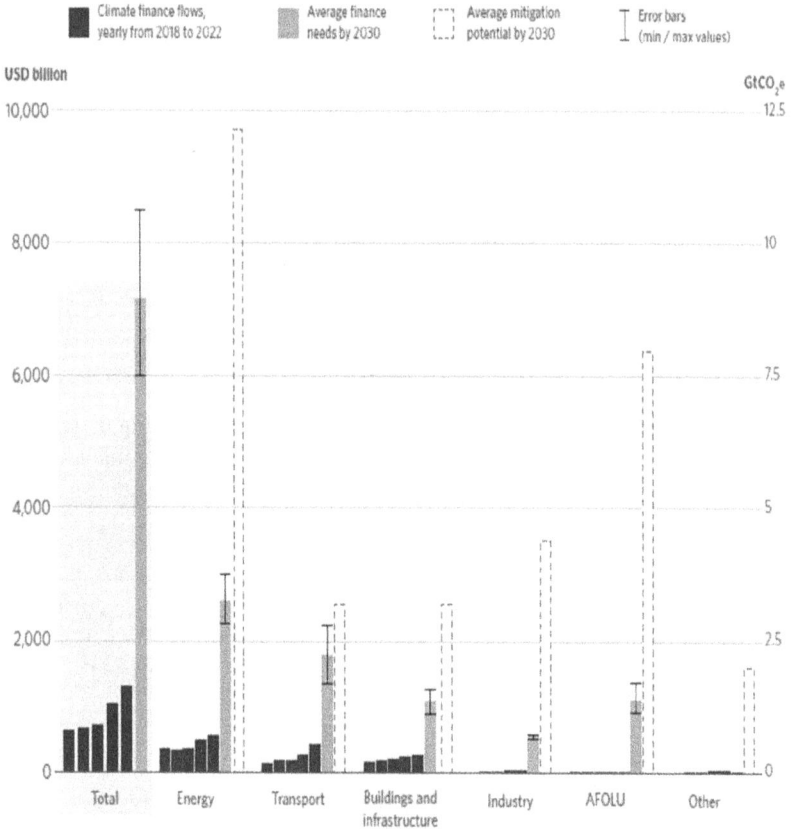

FIGURE 3.2   The historical flows of climate financing (2018–2022) and the demand for climate finance from 2023 to 2030. (Source: CPI, 2024.)

the full breadth of climate finance. This informational deficit impedes effective planning within government ministries, potentially leading to resource duplication and wasteful allocation (Table 3.3). Furthermore, some NGOs' reluctance to provide financial information originates from reasonable concerns, such as greater government engagement, potential diversion of funding through government channels, and the imposition of excessive operational control. This lack of transparency, while acceptable from the standpoint of non-governmental organizations (NGOs), maintains the information vacuum, preventing thorough evaluations and coordinated measures to combat climate change.

TABLE 3.3 Financial management systems for climate expenditures: modification vs. development

NO.	MODIFICATIONS TO EXISTING FINANCIAL MANAGEMENT SYSTEMS	DEVELOPMENT OF A NEW CLIMATE FINANCING TRACKING SYSTEM
1	Merits of existing system modifications: • Integrating climate finance into existing financial plans would improve resource allocation and monitoring, resulting in more effective climate action. • Leveraging staff experience with current systems reduces training expenses, resulting in a quick and effective implementation.	Merits of new tracking system development: • Standardizing data collecting and reporting procedures will make it easier to compare data across countries, allowing for more accurate global analysis and informed policy decisions.
2	Demerits of existing system modifications: • Modifying existing systems provides a trifecta of challenges: high technical complexity, huge financial investment, and lengthy implementation timetables. • Proposed changes may face bureaucratic inertia, especially when significant changes are necessary, thereby impeding or halting progress.	Demerits of new tracking system development: • Implementing such a system needs considerable procedural overhauls and training, which impose enormous technical, temporal, and financial. • Collaborative design across various countries is a difficult coordination challenge, potentially resulting to lengthy negotiations and implementation delays. • Risk of poor synchronization with existing financial monitoring systems, resulting in operational inefficiencies and weakened oversight.

3.6.1 Climate Change Adaptation Project Monitoring and Tracking

Integrating climate change adaptation into development is a difficult balancing act (Omukuti, 2024). While synergistic opportunities exist where adaptation measures directly support existing development goals, such as resilient infrastructure that boosts both economic growth and climate preparation, major financial constraints may occur. Incorporating adaptation into development plans frequently involves significant investments, which may divert resources away from other critical development programs and raise arguments about national priorities (Shirai, 2023; Songwe et al., 2022). To effectively monitor and manage climate change adaptation programs, it is necessary to understand their nuanced link with larger development goals. While synergistic integration can increase beneficial outcomes, adding adaptive techniques may incur large costs, forcing challenging priority judgments. However, foreign support can help to reduce this stress and prevent national resource conflicts. A mainstreaming approach (Table 3.4) to evaluation requires recognizing climate change as a development concern, connecting human well-being to environmental factors in a variety of sectors such as agriculture and health (Arezki, 2021; Chirambo, 2017). Assessing existing climatic extremes and their developmental consequences provides a framework for forecasting future climate threats. Combining climate effect data with development statistics is critical, but data limits and inherent uncertainties present significant methodological hurdles, complicating the whole review process.

The "Development Pathways" approach focuses on the complex interplay between climate change consequences and adaptation options, which ultimately shapes sustainable development outcomes. Climate change consequences, which include extreme weather events, resource constraints, and health concerns, play an important role in affecting development trajectory. Simultaneously, adaptation techniques such as investments in resilient infrastructure, strong policy frameworks, and adequate social safety nets help to offset these effects. The interaction of these inputs in the "Development Pathways" box determines the ensuing sustainable development outcomes, which include economic growth, social equity, and environmental sustainability (Figure 3.3). However, the framework's effectiveness is dependent on several key considerations: context specificity, which ensures that adaptation strategies are tailored to unique regional challenges; equity and justice, which ensures that vulnerable populations are prioritized and benefits are distributed fairly; and integration and mainstreaming, which embeds climate change considerations into all aspects of development planning to foster a holistic and resilient approach.

3.6.2 Climate Change Mitigation Projects Monitoring and Tracking

The urgent need to combat climate change has pushed forestry projects to the forefront of worldwide efforts to minimize GHG emissions (Kellogg, 2019; Ogato et al., 2009). However, the success of these programs is dependent on strong monitoring and assessment methods. Without proper tracking of their

TABLE 3.4 Climate mainstreaming objectives and checklist

NO.	TARGETS	EXPECTED RESULTS	INDICATORS	RISKS AND ASSUMPTIONS
1	**Food, Nutrition and Security**	• Productivity enhanced, • Climate-smart farming practices or techniques implemented, • Food security increased for vulnerable households, • Climate change adaptation proactively mainstreamed in the sector.	• % Increase in yield per hectare (tons) • % Increase in agro bio • % of targeted population adopting one or more climate-smart practices • % of targeted food secure population	• Useful and pertinent climate data to enable planning (observations, forecasts, longer-term projections). • Small-scale adaptation strategies that focus innovation.
2	**Water**	• WRD/Investments in water take into account potential effect of climate change, • Access to safe drinking water became more reliable, during times of extreme climatic stress and high demand, • Prioritizing sustainable natural resource management (SNRM) in landscapes at risk, • The kind and scope of anticipated climate change consequences are considered in SNRM planning, • Rangelands and watersheds were more resilient.	• # of people receiving potable water from new or rehabilitated systems and sources • % of targeted population with year- round access to safe drinking water • # of SNRM plans integrating adaptation developed for vulnerable landscapes • # of hectares of vulnerable landscapes where SNRM plans integrating adaptation	• Mutual cooperation and community involvement among all parties, • At the right scale and useful formats, pertinent climate information, • Establishment of ecosystem services valuation, • Enable adaptation planning in natural resource sector, usable climatic information is provided.

(continued)

TABLE 3.4 (Continued)

NO.	TARGETS	EXPECTED RESULTS	INDICATORS	RISKS AND ASSUMPTIONS
3	**Health**	• Protocols and systems for environmental and health surveillance have been improved to account for information on climate change's impact on health parameters, • An increased understanding of how climate change will affect health.	• % of target population covered by environment and health surveillance systems integrating climate change • % of health workers aware of climate change impacts	• Institutional framework and integrated infrastructure for human health, • Access to helpful forms and adequate scales of pertinent climate information enable adaptation planning.
4	**Gender**	• Technology requirements evaluations for disadvantaged sectors or groups • Development and use of pertinent adaptation technologies	• # of people (women) benefitting from adaptation technologies	• Integration methods for domestic and imported technologies.
5	**Job, Skill and work**	• Analyzing the information and knowledge requirements for adaptation and developing pertinent research plans, • Increased funding for and/or accessibility to research that focuses on adaptation and/or takes climate change impacts into account • Evidence of adoption of findings from studies on climate change adaptation	• # of research products produced that address climate change adaptation/consider climate change impacts • Documented cases of research informing climate-adaptive policies and practices	• Sufficient funds available • Cooperative relationships between research, academic, training, and implementation institutions.

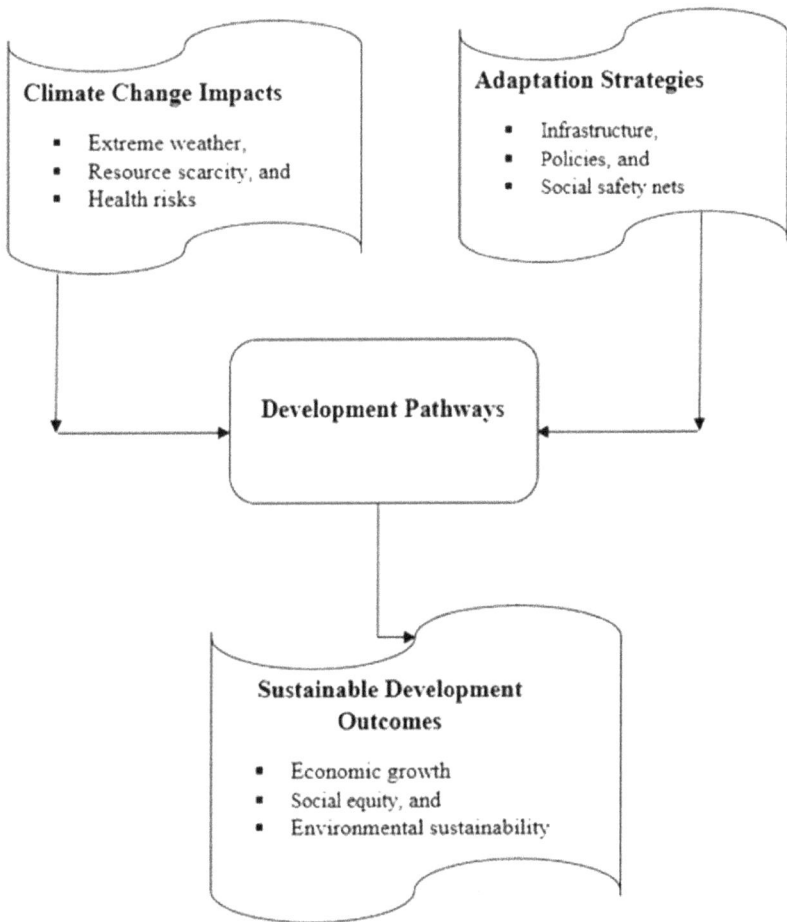

FIGURE 3.3 Climate change and adaptation development pathways.

impact on GHG emissions and other essential qualities, the expected climate benefits may be unmet, compromising both global climate protection and national compliance with international accords (Arezki, 2021; Bannor, 2022; Labatt & White, 2011). As a result, a rigorous framework for monitoring, evaluation, reporting, verification, and certification (MERVC) is required to assure the integrity and transparency of forestry-based climate mitigation projects. Recent changes to MERVC rules have tried to provide a comprehensive roadmap for project developers, evaluators, verifiers, and certifiers. These guidelines address the various issues that come with estimating the climate benefits of forestry initiatives, which include operations like afforestation, reforestation, and averted deforestation. A key challenge is accurately estimating both gross and net carbon reductions, which necessitates rigorous assessment of carbon stock changes across various forest carbon pools. This entails using appropriate approaches, such as field surveys, remote sensing, and modeling, to account for uncertainties and ensure the accuracy of carbon accounting.

The guidelines also underline the necessity of considering aspects other than carbon sequestration, including biodiversity protection, societal benefits, and ecosystem services. Integrating these co-benefits into the MERVC framework improves the overall assessment of forestry projects and supports sustainable development. These guidelines aim to improve the reliability and comparability of forestry initiatives by establishing clear and consistent methods for data collection, analysis, and reporting, promoting better confidence among stakeholders and easing climate financing mobilization. Finally, the successful deployment of effective MERVC systems is critical to ensure that forestry projects provide concrete and verifiable benefits to climate change mitigation.

3.6.3 Public and Private Climate Finance Projects Tracking and Follow-Ups

Green budget tagging (GBT) is an important public financial management strategy used to improve transparency and accountability in climate-related spending (Digitemie & Ekemezie, 2024; Rumble & First, 2021). Governments can effectively identify, classify, and weigh their spending and earnings connected to the environment and climate by assigning specific tags or codes to budget lines (Chaudhury, 2020; Martín Casas et al., 2022). This process enables a detailed examination of how public monies are allocated to climate goals, allowing for more informed budgetary planning. The classifications used in GBT varied across countries, reflecting differing priorities and methods. Some countries categorize fiscal measures depending on their influence on reduction or adaptation, while others group them based on their compatibility with certain climate policies. This flexibility enables governments to adjust their GBT systems to their own settings and goals.

The capacity of GBT to evaluate the climate impact of various budget components is critical. This entails determining if each expenditure or revenue item has a positive, neutral, or negative impact on environmental goals. Governments can acquire a thorough knowledge of their budget's total climate impact by assigning a matching tag. This review would be ideal to start with the implementation of a tagging system, ensuring that tracking green expenditures becomes an integral part of the budgeting process. Furthermore, GBT is critical for enhancing budgetary coherence with climate and environmental goals. GBT improves openness and accountability by providing clear and accessible information on climate-related spending, allowing stakeholders to monitor and evaluate the government's progress toward meeting its climate goals. It is critical to distinguish between green budget and climate budget tags. While GBT incorporates a larger variety of environmental factors, climate budget tagging focuses solely on climate-related budget initiatives. Climate budget tagging is a government-led procedure that identifies, measures, and monitors public expenditures on climate change. This tighter emphasis enables a more thorough examination of how public money is used to solve climate concerns. Both GBT and climate budget tagging are critical instruments for encouraging sustainable development and ensuring that public funds are allocated effectively to climate action.

3.7 CONCLUSION

The conclusion argues that a multifaceted approach is required to address the health challenges posed by climate change in Africa and developing countries. Thus, addressing the rising health concerns brought by climate change in Africa requires a comprehensive and multidimensional approach. A key component is the construction of a strong framework, such as the UAE Framework for Global Climate Resilience, to systematically track progress and prioritize the most pressing requirements. This enables a coordinated and effective global response, with resources directed toward impactful solutions. Empowering communities and addressing underlying health inequities through clean water, sanitation, and nutritional food are all critical components of a just and equitable approach. These programs address the vulnerabilities worsened by climate change, developing resilience at the community level. Furthermore, major investment in adaptation financing, specifically for the African health sector, is required. This financial infusion will allow for the improvement of early warning systems, the construction of climate-resilient hospital infrastructure, and the provision of specialized training for healthcare staff to help them handle climate-related health hazards. To complement these adaptation measures, bold national climate change mitigation strategies are required. In this context, Ethiopia's NDCs, which include a four-pronged approach focusing on sustainable agriculture, forest conservation, renewable energy, and energy efficiency, demonstrate a proactive position. Countries that employ these measures hope to obtain a twin benefit: lowering greenhouse gas emissions while also promoting long-term economic development. Finally, successfully navigating climate change's health implications in Africa will require a concerted global effort that combines international support with powerful national initiatives, safeguarding the continent's health and resilience.

REFERENCES

Amare, Z. Y., Geremew, B. B., Kebede, N. M., & Amera, S. G. (2024). Impacts of El Niño-Southern oscillation on rainfall amount and anticipated humanitarian impact. *Environment, Development and Sustainability*, *26*(12), 1–19.

Angerer, L., Dachrodt, M., Gimeno, M. M., & Healy, S. Operationalising the global goal on adaptation.

Arezki, R. (2021). Climate finance for Africa requires overcoming bottlenecks in domestic capacity. *Nature Climate Change*, *11*(11), 888–888.

Asfaw, A., Simane, B., Bantider, A., & Hassen, A. (2019). Determinants in the adoption of climate change adaptation strategies: Evidence from rainfed-dependent smallholder farmers in north-central Ethiopia (Woleka sub-basin). *Environment, Development and Sustainability*, *21*(5), 2535–2565.

Bannor, F. (2022). *Agriculture, climate change and technical efficiency: The case of sub-Saharan Africa*. University of Johannesburg.

Barnard, S., Watson, C., Greenhill, R., Caravani, A., Trujillo, N. C., Hedger, M., & Whitley, S. (2014). *Climate finance: Is it making a difference?* Overseas Development Institute.

Baskaran, G., Ekeruche, A., Heitzig, C., Ordu, A. U., & Senbet, L. W. (2023). Financing climate-resilient infrastructure in Africa.

Bendell, J. (2018). *Deep adaptation: A map for navigating climate tragedy*. University of Cumbria.

Boespflug, C. M. (2024). Assessing water, energy, and food security in Ethiopia: A CLEWs Nexus approach.

Bogale, G. A., & Erena, Z. B. (2022). Drought vulnerability and impacts of climate change on livestock production and productivity in different agro-ecological zones of Ethiopia. *Journal of Applied Animal Research*, *50*(1), 471–489.

Bowman, M., & Minas, S. (2019). Resilience through interlinkage: The green climate fund and climate finance governance. *Climate Policy*, *19*(3), 342–353.

Briera, T., & Lefèvre, J. (2024). Reducing the cost of capital through international climate finance to accelerate the renewable energy transition in developing countries. *Energy Policy*, *188*, 114104.

Butler, C. D. (2018). Climate change, health and existential risks to civilization: A comprehensive review (1989–2013). *International Journal of Environmental Research and Public Health*, *15*(10), 2266.

Chaudhury, A. (2020). Role of intermediaries in shaping climate finance in developing countries—lessons from the green climate fund. *Sustainability*, *12*(14), 5507.

Chirambo, D. (2017). Enhancing climate change resilience through microfinance: Redefining the climate finance paradigm to promote inclusive growth in Africa. *Journal of Developing Societies*, *33*(1), 150–173.

Chirisa, I., Gumbo, T., Gundu-Jakarasi, V. N., Zhakata, W., Karakadzai, T., Dipura, R., & Moyo, T. (2021). Interrogating climate adaptation financing in Zimbabwe: Proposed direction. *Sustainability*, *13*(12), 6517.

CPI. (2024). *Global landscape of climate finance 2024: Insights for COP 29*. Available online: climatepolicyinitiative.org/publication/global-landscape-of-climate-finance-2024

Conway, D., & Schipper, E. L. F. (2011). Adaptation to climate change in Africa: Challenges and opportunities identified from Ethiopia. *Global Environmental Change*, *21*(1), 227–237.

Digitemie, W. N., & Ekemezie, I. O. (2024). Assessing the role of climate finance in supporting developing nations: A comprehensive review. *Finance & Accounting Research Journal*, *6*(3), 408–420.

Ewbank, R., Perez, C., Cornish, H., Worku, M., & Woldetsadik, S. (2019). Building resilience to El Niño-related drought: Experiences in early warning and early action from Nicaragua and Ethiopia. *Disasters*, *43*, S345–S367.

Fleishel, R., Cauthen, C., Daniewicz, S., Baker, A., Jordon, J. B., & TerMaath, S. (2021). Characterization of surface fatigue crack nucleation and microstructurally small crack growth in high strength aluminum alloys. *Frontiers in Materials*, *7*, 590747.

Garrett, L. (2013). *Existential challenges to global health*. Center on International Cooperation.

Gordon, N. J. (2023). Climate finance: An overview. *Environment: Science and Policy for Sustainable Development*, *65*(4), 18–26.

Kellogg, W. W. (2019). *Climate change and society: Consequences of increasing atmospheric carbon dioxide*. Routledge.

Labatt, S., & White, R. R. (2011). *Carbon finance: The financial implications of climate change*. John Wiley & Sons.

Martín Casas, N., & Remalia Sanogo, A. (2022). Climate finance in West Africa: Assessing the state of climate finance in one of the world's regions worst hit by the climate crisis.

Michaelowa, A., Espelage, A., Lieke't Gilde, N. K., Censkowsky, P., Greiner, S., Ahonen, H.-M., ... Dalfiume, S. (2021). Article 6 readiness in updated and second NDCs. *Perspectives climate group & climate focus*, 1–64.

Mikulewicz, M. (2018). Politicizing vulnerability and adaptation: On the need to democratize local responses to climate impacts in developing countries. *Climate and Development*, *10*(1), 18–34.

Møller, V., & Roberts, B. J. (2021). *Quality of life and human well-being in sub-Saharan Africa*. Springer.

Mosadeghrad, A. M., Isfahani, P., Eslambolchi, L., Zahmatkesh, M., & Afshari, M. (2023). Strategies to strengthen a climate-resilient health system: A scoping review. *Globalization and Health*, *19*(1), 62.

Mugambiwa, S., & Kwakwa, M. (2022). Multilateral climate change financing in the developing world: Challenges and opportunities for Africa. *International Journal of Research in Business and Social Science (2147–4478)*, *11*(9), 306–312.

Njuguna, L., Uri, I., & Beauchamp, E. (2024). National monitoring, evaluation, and learning systems for climate change adaptation.

Odhong', C., Wilkes, A., van Dijk, S., Vorlaufer, M., Ndonga, S., Sing'ora, B., & Kenyanito, L. (2019). Financing large-scale mitigation by smallholder farmers: What roles for public climate finance? *Frontiers in Sustainable Food Systems, 3*, 3.

Ogato, G. S., Boon, E. K., & Subramani, J. (2009). Improving access to productive resources and agricultural services through gender empowerment: A case study of three rural communities in Ambo District, Ethiopia. *Journal of human Ecology, 27*(2), 85–100.

Oladosu, A. O., & Chanimbe, T. (2024). A two-pronged approach to understanding reciprocity and mental health relationship in developing countries: Evidence from young informal construction workers in Nigeria. *BMC Public Health, 24*(1), 1851.

Omukuti, J. (2024). The need for a climate-resilient development-aligned framing of innovative climate finance. *Current Opinion in Environmental Sustainability, 66*, 101400.

Rumble, O., & First, J. (2021). Accelerating private sector climate finance in Africa.

Savvidou, G., Atteridge, A., Omari-Motsumi, K., & Trisos, C. H. (2021). Quantifying international public finance for climate change adaptation in Africa. *Climate Policy, 21*(8), 1020–1036.

Scott, A., Worrall, L., & Patel, S. (2018). Aligning energy development and climate objectives in nationally determined contributions. Retrieved on January, 17, 2019.

Scullion, J. J., Vogt, K. A., Drahota, B., Winkler-Schor, S., & Lyons, M. (2019). Conserving the last great forests: A meta-analysis review of the drivers of intact forest loss and the strategies and policies to save them. *Frontiers in Forests and Global Change, 62*, 1–12.

Seltenrich, N. (2018). Safe from the storm: Creating climate-resilient health care facilities. *Environmental Health Perspectives, 126*(10), 102001.

Shirai, S. (2023). *Global climate challenges, innovative finance, and green central banking*. Asian Development Bank.

Shkabatur, J., Bar-El, R., & Schwartz, D. (2022). Innovation and entrepreneurship for sustainable development: Lessons from Ethiopia. *Progress in Planning, 160*, 100599.

Songwe, V., Stern, N., & Bhattacharya, A. (2022). *Finance for climate action: Scaling up investment for climate and development*. Grantham Research Institute on Climate Change and the Environment, London School of Economics and Political Science.

Stenchikov, G. (2021). The role of volcanic activity in climate and global changes. In *Climate change* (pp. 607–643). Elsevier.

Stender, F., Moslener, U., & Pauw, W. P. (2020). More than money: Does climate finance support capacity building? *Applied Economics Letters, 27*(15), 1247–1251.

Teweldebrhan, M., & Dinka, M. (2024). The impact of climate change on the development of water resources. *Global Journal of Environmental Science and Management, 10*(3), 1359–1370.

Timilsina, G. R. (2021). Financing climate change adaptation: International initiatives. *Sustainability, 13*(12), 6515.

Winkler, H., Watson, C., & Bhandari, P. (2024a). Connecting global stocktake outcomes and COP28 workstreams.

Winkler, H., Watson, C., & Bhandari, P. (2024b). Connecting GST outcomes and COP28 workstreams.

The Capacity of Climate Financing in Africa

4

4.1 INTRODUCTION

The studies show that Africa, a continent, which bears a disproportionate weight from the effects of climate change, is aggressively seeking mitigation solutions (Mekonnen, 2014; Strzepek & McCluskey, 2007). The successful implementation of ambitious NDCs, which are dependent on significant financial resources, is central to this effort. Despite international agreements to provide 80% of Africa's climate action financing, the reality on the ground is quite different (Digitemie & Ekemezie, 2024; Khan et al., 2021; Scott et al., 2018). Thus, complex proposal requirements and limited access to these funds have resulted in a significant shortage, emphasizing the urgent need for new measures to overcome the finance imbalance. In this context, the CERF program addresses this essential issue by focusing on capacity development, providing African countries with the technical competence and proposal development abilities needed to negotiate the complex world of climate funding (Chirambo, 2014; Khan et al., 2021). CERF lays the way for a more sustainable and affluent future by empowering these nations to actively participate in climate change mitigation, rather than simply receiving aid (Bayat-Renoux & Glemarec, 2014; Longhurst & Slater, 2022; Swithern, 2021). This initiative, despite its limitations, is critical for building Africa's resilience to climate change and assuring long-term success in fighting its effects.

DOI:10.1201/9781003644378-4

4.2 CLIMATE RESPONSE FACILITY IN AFRICA

Africa, although being disproportionately affected by climate change and actively pursuing mitigation through ambitious NDCs, has a severe bottleneck in obtaining adequate financial resources (Belianska et al., 2022; Tamasiga et al., 2023). Despite pledges for substantial international assistance, which are frequently projected to cover 80% of required climate change financing, onerous proposal criteria and limited accessibility result in a major shortfall (Chirambo, 2014; Rumble & First, 2021). This financial shortage is worsened by capacity restrictions, which prevent African countries from successfully navigating the complicated climate finance landscape. The Climate Empowerment and Resilience Fund (CERF) is developing as a potential option, with a focus on capacity building to provide African governments with the requisite technical competence and proposal development abilities (Longhurst & Slater, 2022; Swithern, 2021). CERF strives to bridge the gap by empowering these countries to actively participate in climate finance, allowing them to invest in climate-resilient development and create a more sustainable future.

4.3 THE OPPORTUNITIES AND CHALLENGES

The CERF presents considerable opportunity for African countries to meet their NDCs by addressing the major finance gap that impedes climate action (Longhurst & Slater, 2022; Weikmans et al., 2021). A significant potential is to increase the ministries of finance's ability to navigate and receive international climate finance (Rumble & First, 2021; Weischer et al., 2016). By providing training and technical help, CERF enables these ministries to create bankable initiatives that meet the criteria of global climate funds. This capacity building not only helps to facilitate the flow of critical resources, but it also promotes long-term sustainability by preparing countries to acquire future funding on their own. Furthermore, CERF enables the construction of sophisticated climate financing tracking systems, which improves transparency and accountability while also directing funding to priority areas such as climate-smart agriculture, renewable energy, and sustainable land management. However, considerable obstacles persist. African countries are vulnerable to global financial volatility due to their reliance on international assistance, which is expected to account for 80% of total funding needs (Chang et al., 2019). Current public and private sector cash flows fall far short of the necessary levels, resulting in a significant funding gap (Arezki, 2021; Rumble & First, 2021). Navigating the complicated terrain of international climate funding, including satisfying tight qualifying criteria and producing attractive bids, necessitates extensive knowledge and resources (Colenbrander et al., 2018; Gordon, 2023). Furthermore, while government-to-government cooperation is encouraged, achieving effective knowledge sharing and coordinated action across varied national contexts can be challenging. Maintaining momentum beyond initial funding acquisition, as well as guaranteeing long-term impact through good NDC implementation and monitoring, is a key challenge that necessitates ongoing capacity building and institutional improvement. The study indicate that the least developed countries (LDCs) and small island developing states (SIDS) face significant barriers to accessing public climate finance due to high debt burdens and strained government budgets, while private finance contributions remain critically low (Figure 4.1), consistently falling below 10% for LDCs (except in 2021) and below 2% for low- and lower middle-income SIDS between 2018 and 2022 (CPI, 2024).

Countries are recognizing the importance of developing strong technical capacity in order to fully realize the potential of climate finance (Arezki, 2021; Digitemie & Ekemezie, 2024; Weischer et al., 2016). This emphasis on capacity building for GCF access modalities aims to enable countries to effectively mobilize resources for climate action (Bayat-Renoux & Glemarec, 2014; Briera & Lefèvre, 2024). Here is a breakdown of some major areas of focus:

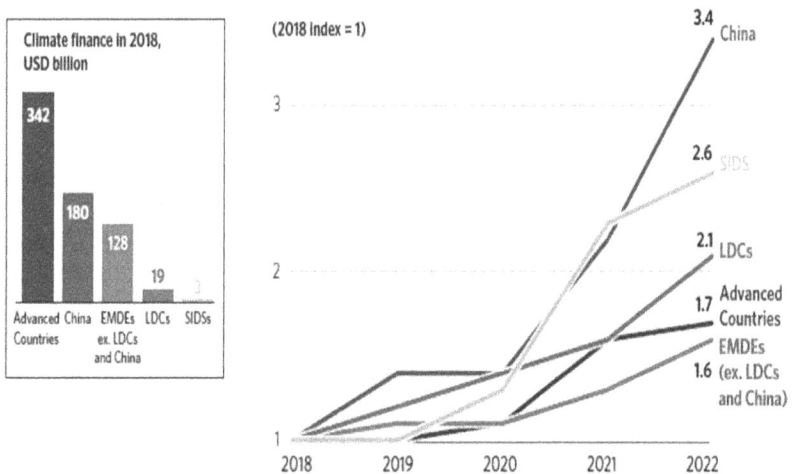

FIGURE 4.1   The comparison of countries' growth in climate financing (right) and absolute values (left). (Source: CPI, 2024.)

A. **Establishing a Project Pipeline for Climate Financing Instruments:** A well-defined project pipeline ensures a consistent flow of high-quality proposals that align with GCF priorities (Bowman & Minas, 2019). This includes identifying potential projects, creating concept notes, and preparing full funding proposals that meet the GCF's stringent criteria. Building this capacity enables countries to make compelling cases for securing critical climate finance (Chaudhury, 2020; Doku et al., 2021b).

B. **Managing Environmental and Social Safeguards for Climate-Focused Projects:** The GCF prioritizes projects that provide climate benefits while reducing environmental and social risks (Chaudhury, 2020). Building capacity in this area enables countries to assess the potential environmental and social impacts of projects, develop mitigation strategies, and ensure compliance with GCF safeguard policies (Bowman & Minas, 2019; Chaudhury, 2020). This encourages responsible project development while protecting vulnerable communities.

C. **Conducting Economic and Financial Analysis for GCF Projects:** Strong economic and financial analysis is required to obtain GCF approval. Building this capacity enables countries to accurately assess project costs, benefits, and risks (Doku et al., 2021b). This includes demonstrating cost-effectiveness, the potential for long-term financial sustainability, and obtaining co-financing from alternative sources. Strong financial analysis improves a project's competitiveness and ensures that funds are allocated efficiently.

D. **Implementing Climate Budget Tagging and Tracking Systems:** Effective tracking of climate finance enables countries to demonstrate transparency and accountability in resource allocation (Chaudhury, 2020; Doku et al., 2021b). Building capacity in this area entails putting in place systems to identify budgetary allocations designated for climate action and to track the flow and impact of these funds. This transparency builds trust among donors and ensures that climate finance meets its objectives.

E. **Identifying and Gaining Access to Non-traditional Climate Finance Sources, such as Debt-for-climate Swaps:** Beyond the GCF, a growing number of novel financing mechanisms are emerging (Bowman & Minas, 2019; Chaudhury, 2020). Building capacity in this area enables countries to consider options such as debt-for-climate swaps, which forgive a portion of a country's external debt in exchange for investments in climate projects (Doku et al., 2021b; Savvidou et al., 2021). This diversification of funding sources has the potential to significantly increase the available resources for climate action.

Countries can improve their access to and mobilization of climate finance by focusing on these key areas of capacity building (Doku et al., 2021a; Rumble & First, 2021). This gives them the ability to implement effective climate change mitigation and adaptation strategies, accelerating progress toward a more sustainable future.

4.4 CLIMATE FINANCING IN AFRICA

4.4.1 Barriers to Climate Fund Access: Proposal Issues for Africa's Development

Access to critical climate financing remains a significant challenge for African countries (Baskaran et al., 2023; Odhong' et al., 2019; Stender et al., 2020). A significant barrier is a lack of well-developed proposals that meet the stringent requirements of international climate funds (Colenbrander et al., 2018; Savvidou et al., 2021). This translates into a few key capacity limitations:

- **Proposal Expertise:** Creating compelling proposals for funds such as the GCF and the Loss and Damage Funds requires specialized knowledge. African countries may struggle to effectively write and report on projects in accordance with these institutions' guidelines.
- **Financial Analysis Shortfalls:** Securing funds frequently relies on thorough economic and financial analyses. However, limitations in conducting such analyses may prevent African countries from demonstrating the cost-effectiveness and long-term benefits of their climate projects.
- **Project Management Capacity Gap:** Effectively managing climate projects necessitates skilled individuals at both the national and local levels. A lack of training and capacity building in project management can impede successful implementation.
- **Data and Communication Shortcomings:** Transparent reporting and communication are critical for obtaining and managing climate funds. Inadequate IT infrastructure and expertise can make it difficult for African countries to implement effective reporting and communication systems.
- **Budget Tracking Obstacles:** Climate-related expenditures must be monitored and tracked in order to ensure accountability and make future funding decisions. The lack of a reliable climate budget

tagging and tracking system can impede effective resource allocation in African countries.

- **Untapped Funding Sources:** Aside from traditional funding sources, there is a potential wealth of private and innovative financing mechanisms. However, a lack of awareness and expertise in mapping and gaining access to these non-traditional avenues can limit African countries' overall resource base.

Addressing these capacity gaps is crucial to unlock the potential of climate financing for Africa (Briera & Lefèvre, 2024; Savvidou et al., 2021; Stender et al., 2020). By investing in proposal development skills, financial analysis training, project management expertise, and robust data management systems, African countries can overcome these hurdles and access the resources needed to tackle climate change and build a more sustainable future (Baskaran et al., 2023; Mungai et al., 2021; Nakhooda et al., 2011).

4.4.2 Climate Financed Project Application Guides

Climate finance project application guides are crucial blueprints for navigating the complex environment of climate action, transforming abstract global goals into concrete local actions. They go beyond simple procedural instructions, serving as strategic frameworks for successfully channeling financial resources toward projects that improve resilience and minimize climate threats. These instructions promote transparency and accountability in climate finance mechanisms by defining eligibility criteria, application processes, and reporting requirements. Notably, developed climate finance (DCF) emerges as a critical option, advocating for the decentralization of climate spending by empowering local communities and subnational governments. This localized model recognizes their particular awareness of regional concerns, allowing them to establish subnational climate funds and prioritize and implement context-specific climate initiatives.

Furthermore, DCF goes beyond financial delivery, focusing on the creation of specific instruments for climate risk assessment, planning, and monitoring and evaluation. These technologies provide local players with the essential data and analytical frameworks to make educated investment decisions, while rigorous monitoring and evaluation ensures ongoing project improvement and resource efficiency. Finally, climate finance project application guides, particularly those supporting DCF, play an important role in promoting a sustainable and resilient future by empowering local residents to lead the battle against climate change.

4.4.3 Structures and Procedures for Climate Finance Project Applications

The landscape of climate finance project applications is highly variable, with each country's demands and geographical factors dictating the approach. Since the Clean Development Mechanism (CDM)'s establishment in 2004, there has been a significant expansion, with a pipeline of thousands of projects delivering billions of certified emission reductions. CDM programs have boosted investment in low-carbon technologies, particularly in the energy sector. However, the distribution of these benefits is uneven, with regions such as Africa having difficulty in realizing their full potential. Africa, although having significant reduction potential, has struggled to effectively use the CDM due to inadequate institutional structures (Table 4.1).

Analysis demonstrates a negative association between investment costs and African participation: sectors with high investment costs have lower African involvement, whereas those with lower investment per CER have higher African engagement (Table 4.1). This gap demonstrates the importance of financial accessibility and institutional capacity in influencing the effectiveness of climate financing efforts. Intermediaries play an important role in closing these gaps, and continuing research is looking into different propositions and learning models to improve climate action in developing countries.

Several important concepts are under development that seek to improve the Green Climate Fund's (GCF) strategy. One important discovery is the difference in project approval durations between national and international Accredited Entities (AEs). National AEs frequently face much longer approval processes, causing projects to become outdated and more expensive. This lag can inhibit innovation and impede the timely implementation of critical climate policies. Furthermore, because national AEs execute projects independently, other local intermediaries are less exposed to the GCF process. This isolation can stymie knowledge sharing and capacity creation, preventing the growth of a healthy and varied climate finance ecosystem. Finally, national AEs often propose initiatives that correspond with their current expertise and skills, which, while efficient, may limit the breadth of national climate action. This propensity might lead to a fragmented approach, perhaps missing key areas that demand immediate attention. Testing and improving these propositions is critical to increasing the GCF's effectiveness and accelerating global climate actions (Table 4.1).

TABLE 4.1   Structures and procedures for climate finance project applications

ASPECT OF CLIMATE FINANCE PROJECT APPLICATIONS	KEY OBSERVATIONS/ CHALLENGES	IMPLICATIONS/ RECOMMENDATIONS
Clean Development Mechanism (CDM) Experience	• Significant expansion since 2004, with numerous projects and billions of CERs. • Uneven distribution of benefits, with Africa struggling to realize full potential. • Negative association between investment costs and African participation.	• Highlight the importance of financial accessibility and institutional capacity. • Address the gap in benefit distribution, particularly for regions like Africa. • Recognize the role of intermediaries in bridging these gaps.
Green Climate Fund (GCF) Project Approval	• National Accredited Entities (AEs) face longer approval durations compared to international AEs. • Longer durations can lead to projects becoming outdated and more expensive. • National AEs' independent execution limits exposure of other local intermediaries to the GCF process. • National AEs tend to propose projects that are in line with their current expertise.	• Streamline approval processes for national AEs to ensure timely implementation. • Promote knowledge sharing and capacity building among local intermediaries. • Encourage a broader range of project proposals to address diverse climate action needs. • Improve the GCF strategy to increase effectiveness.
General Landscape	• Landscape is highly variable, dictated by country demands and geographical factors.	• Customized approaches are required for each country.

4.4.4 Building Capacity for Effective Climate Finance in Africa

Climate change represents a significant threat to Africa's development (Doku et al., 2021a). Unlocking effective climate financing is critical in addressing this challenge (Baskaran et al., 2023; Mungai et al., 2022). However, simply having access to funds is not sufficient (Colenbrander et al., 2018; Doku et al.,

2021a). There is a critical need to increase African nations' capacity to effect-ively implement these financial resources. Here's how training and strategic partnerships can play an important role:

a. **Empowering Action: Training in Directives and Manuals**
   Clear direction and well-defined procedures are critical to successful climate finance implementation (Arezki, 2021; Doku et al., 2021a). This is where training in creating directives and manuals comes into play. By providing individuals and institutions with the skills required to develop these resources, Africa can ensure:
   - **Transparency and accountability:** Clear directives and manuals promote open financial management practices. This builds trust with international partners and ensures funds are used for their intended purpose.
   - **Streamlined Processes:** Standardized procedures outlined in manuals facilitate project implementation. This reduces delays and increases the effectiveness of climate finance initiatives.
   - **Capacity Building:** The process of creating directives and manuals is a valuable learning opportunity. It promotes a better understanding of climate finance mechanisms and encourages local actors to take more active roles (Arezki, 2021; Doku et al., 2021a).

b. **Unlocking Potential through Public-Private Partnerships (PPPs) and Forest Carbon Financing**
   PPPs are an effective tool for mobilizing additional financial resources for climate action in Africa (Briera & Lefèvre, 2024; Rumble & First, 2021). Training programs can provide both public and private sector actors with the knowledge and skills required to:
   - **Develop bankable projects:** By understanding the needs of investors and international donors, partnerships can create projects that are appealing for funding.
   - **Navigate legal frameworks:** A clear understanding of the legal frameworks governing PPPs and forest carbon financing enables the creation of compliant and sustainable projects.
   - **Share risks and rewards:** Training can assist in establishing effective mechanisms for sharing the risks and rewards associated with climate finance initiatives, fostering a sense of shared responsibility and long-term success (Doku et al., 2021a).

c. **Involving the Private Sector and Sharing Benefits**
   The private sector plays a critical role in promoting climate-resilient development in Africa (Doku et al., 2021a; Nakhooda et al., 2011). Training programs that focus on private sector participation in carbon financing can provide businesses with the knowledge to:

- **Identify and assess carbon reduction opportunities:** Businesses can learn how to identify and implement projects that reduce their carbon footprint, increasing their appeal to carbon credit markets.
- **Create innovative financing solutions:** Training can encourage the creation of novel financing mechanisms that leverage private sector investment for climate action (Doku et al., 2021a).
- **Create fair and equitable benefit-sharing mechanisms:** Training can assist in the development of transparent systems for sharing the benefits of carbon reduction projects with local communities, ensuring a just transition to a low-carbon economy (Briera & Lefèvre, 2024).

African countries can maximize the potential of climate finance by investing in training and forming strategic partnerships (Briera & Lefèvre, 2024; Colenbrander et al., 2018). This will not only help to mitigate the effects of climate change, but will also contribute to the continent's long-term viability and prosperity.

4.4.5 Guides and Support for Successful Climate Negotiations

Successful climate negotiations require strong guidelines and support structures, particularly for developing countries and Africa, which are disproportionately affected by climate change. Understanding the dynamics of these agreements necessitates a strong theoretical base, such as regime theory, which analyzes the political economy of climate change management. This concept highlights Africa and poor countries' structural disadvantages within the international climate regime, emphasizing their marginalization. To address this, a robust continental framework supported by endogenous resources is critical for improving the global climate architecture. Africa has been depicted as a vulnerable victim of climate change, with an emphasis on adaptation. However, the debate over climate change has evolved dramatically. Initially motivated by environmental concerns at COP1 in 1995, it has evolved into a fundamental socioeconomic discussion, as evidenced at the 2002 World Summit on Sustainable Development. This gathering, honoring the Rio gathering, highlighted the importance of combating climate change through the UNFCCC. The Kyoto Protocol, enacted at COP3, underlined the importance of engaging the global industrial sector in order to reduce the effects of climate change.

Africa has played a critical role in developing international climate agreements. African nations prioritized climate finance during COP26 in

Glasgow, advocating for the $100 billion per year pledge made at COP15 in Copenhagen. To strengthen their negotiation position, Africa must improve its ability to absorb this money effectively. This includes establishing feasible initiatives, boosting industrial value chains, building a strong private sector, and improving skill levels. Africa and developing countries are committed to effective climate negotiations at future COPs, particularly COP29. Their ongoing commitment and advocacy for equitable and effective climate action are critical to the long-term success of international climate agreements. As a result, providing detailed guidelines and support for these negotiations is critical in ensuring that developing countries can fully engage and contribute to a sustainable global future.

4.5 THE MILESTONES MADE AND THE HINDRANCES

The UNFCCC Secretariat has issued comprehensive reports analyzing countries' NDCs and LT-LEDS submitted under the Paris Agreement in 2015 (Kissinger et al., 2019; Weikmans et al., 2021). These reports offer a sobering assessment of the current state of global climate action, emphasizing the critical need for increased efforts to mitigate climate change.

Despite some progress by individual countries, synthesis reports show that current national climate plans fall short of the ambitious goals outlined in the Paris Agreement (Khan et al., 2021; Weikmans et al., 2021). The global temperature rise is still expected to exceed 1.5°C, a critical threshold above which the worst effects of climate change become more likely (Khan et al., 2021; Kissinger et al., 2019). The reports emphasize that a more rapid and significant reduction in greenhouse gas emissions is required to avoid these catastrophic consequences. Even though 2030 emissions are expected to be slightly lower than 2019 levels, this drop is not enough to meet the scientific imperative of reaching a peak in emissions before 2030. The reports argue that achieving this goal necessitates the full implementation of conditional elements in the NDCs (Kissinger et al., 2019; Weikmans et al., 2021). However, this implementation is heavily dependent on factors such as increased financial support, technology transfer, technical cooperation, capacity building, and the proper operation of market-based mechanisms. On the other hand, the UNFCCC synthesis reports highlight the critical need for increased international cooperation and accelerated climate action (Winkler et al., 2024a). To avoid the most severe consequences of climate change, countries must significantly strengthen their

national climate plans, mobilize adequate resources, and encourage collabora-
tive efforts to address this global challenge.

4.5.1 COP28: A Mixed Bag of Financial Commitments

The first contact group on long-term climate finance (LTF) at COP28 in Dubai
in 2023 kicked off discussions on a critical aspect of climate action (Winkler
et al., 2024b). During this meeting, various parties presented their expectations
and key components for a draft decision. A major point of contention arose over
the inclusion of progress toward the USD 100 billion goal set by developed
countries.

Developed countries advocated for the inclusion of the most recent
Organisation for Economic Co-operation and Development (OECD) report,
which indicates that this goal may have been met in 2023. However, developing
countries argued that such conjectures were inappropriate and that the emphasis
should be on concrete numbers and verifiable progress (Winkler et al., 2024a).
Egypt, representing the Africa Group, emphasized that the LTF's purpose
goes beyond simply tracking whether developed countries are meeting their
pledges. It emphasized the importance of assessing existing gaps and building
on the discussions from last year's high-level roundtable in Sharm el-Sheikh.
Egypt also urged that the decision address issues of transparency and methodology in accounting for climate finance. It warned against focusing on trends
and expectations, stating that the decision should be based primarily on tangible achievements. Furthermore, Egypt brought up the critical issue of debt
and the need for equitable burden sharing among developed countries in light
of the USD 100 billion goal.

The recently concluded COP28 summit in Dubai made some progress
toward securing financial commitments for climate action, particularly in Africa
(Winkler et al., 2024b). However, the funds raised fall short of the massive
investments needed to address the continent's vulnerability to climate change.
One notable accomplishment was the establishment of the Loss and Damage
Fund, which seeks to compensate developing countries for the irreversible
effects of climate change. Africa, which is especially vulnerable to these effects,
has long advocated for such a fund. While the initial $792 million pledge is a
step in the right direction, it is insufficient to meet the continent's requirements.
In addition to the Loss and Damage Fund, an unprecedented $85 billion was
raised for global climate action. This includes pledges to support renewable
energy, energy efficiency, and other climate-related initiatives (Figure 4.2).

The Global Renewables and Energy Efficiency Pledge, which was
launched at COP28, has the potential to help Africa transition to cleaner energy

LANDSCAPE OF CLIMATE FINANCE IN 2021/2022

Global climate finance flows along their life cycle in 2021 and 2022. Values are averages of two years' data to smooth out fluctuations, in USD billions

SOURCES AND INTERMEDIARIES
Which type of organizations are sources or intermediaries of capital for climate finance?

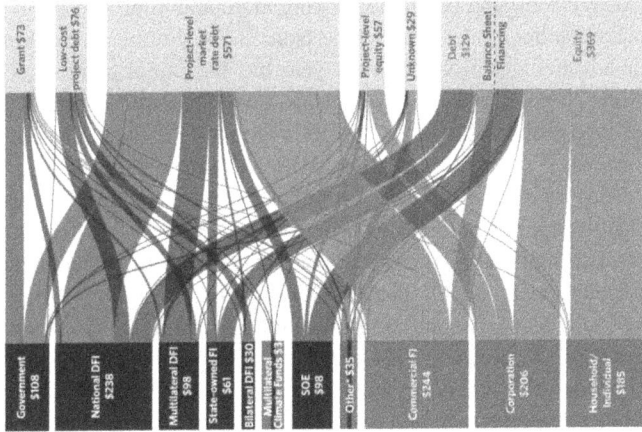

INSTRUMENTS
What mix of financial instruments is used?

USES
What types of activities are financed?

DESTINATION
Where are the flows directed by region?

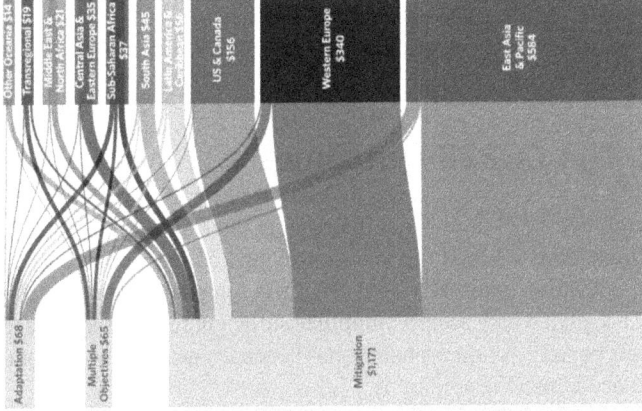

1.3 TRILLION USD ANNUAL AVERAGE

Sources and Intermediaries:
Government $108
National DFI $238
Multilateral DFI $98
State-owned FI $61
Bilateral DFI $30
Multilateral Climate Funds $3
SOE $98
Other $35
Commercial FI $244
Corporation $206
Household/Individual $105

Instruments:
Grant $73
Low-cost project debt $76
Project-level market rate debt $571
Project-level equity $57
Unknown $29
Debt $129
Balance Sheet Financing
Equity $369

Uses:
Adaptation $68
Multiple Objectives $65
Mitigation $1171

Destination:
Other Oceania $14
Transregional $19
Middle East & North Africa $21
Central Asia & Eastern Europe $35
Sub-Saharan Africa $37
South Asia $45
Latin America & Caribbean $60
US & Canada $156
Western Europe $340
East Asia & Pacific $584

PRIVATE   PUBLIC

"Other" public sources include export credit agencies and unknown public funds
"Other" private sources include institutional investors, funds, and unknown

FIGURE 4.2 The climate finance landscape for COP29 preparation. (Source: CPI, 2024.)

sources and reduce emissions significantly (Colenbrander et al., 2018; Winkler et al., 2024a). Similarly, the Global Methane Pledge, which seeks to reduce methane emissions, can aid Africa's efforts to combat climate change. Overall, while COP28 produced some positive results, it is clear that much more work must be done to close the financial gap for climate action, particularly in Africa (Colenbrander et al., 2024). The continent needs significant investments to adapt to the effects of climate change and transition to a low-carbon economy.

4.5.2 The COP28 Outcomes

The recent high-level ministerial dialogue at the Dubai climate talks revealed a significant gap in views between developed and developing countries on the new collective quantified goal (NCQG) for climate finance (Robertson, 2024). This goal, which seeks to mobilize sufficient funds to address the urgent challenges posed by climate change, has become a major point of contention in international climate negotiations.

Developed and developing countries have different perspectives on the NCQG's sources, timeframe, quality, and quantity (Colenbrander et al., 2024). Developed countries frequently advocate for a focus on private sector financing and market-based mechanisms, whereas developing countries emphasize the importance of increased public sector contributions and concessionary financing. Disagreements also persist about the appropriate timeframe for achieving NCQG, with developing countries seeking more ambitious targets in the short term (Colenbrander et al., 2024; Robertson, 2024). Furthermore, the mode of work for advancing the NCQG next year has become a point of contention (Robertson, 2024). Developing countries have expressed a preference for a technical-level approach, which entails negotiating a detailed text to guide future progress (Colenbrander et al., 2018). Developed countries, on the other hand, have advocated for increased political direction and ministerial engagement. This divergence reflects the two groups of countries' different priorities and interests, complicating the process of reaching an agreement on the NCQG (Colenbrander et al., 2024). The NCQG discussions highlight the complexities and challenges of mobilizing the financial resources needed to combat climate change (Robertson, 2024). Bridging the gap between developed and developing countries will necessitate a concerted effort to find common ground and create a comprehensive and equitable approach to climate finance.

Clean Cooking Solutions in Africa:
The most significant outcomes of COP28 were the commitment by world leaders and the African Development Bank to increase funding for clean

cooking solutions in Africa (Colenbrander et al., 2024; Winkler et al., 2024a). This is an important step toward improving public health and livelihoods across the continent. Traditional cooking methods frequently use polluting fuels such as charcoal and wood, contributing to indoor air pollution, respiratory diseases, and deforestation. Millions of people can benefit from better health, cleaner environments, and economic opportunities by switching to cleaner alternatives like solar-powered stoves or biogas.

The Start of the End for Fossil Fuels:
The COP28 agreement's inclusion of a commitment to "transitioning away from fossil fuels" is an important milestone, indicating a potential turning point in the global energy narrative (Colenbrander et al., 2024; Winkler et al., 2024a). While this recognition is an important step toward a sustainable future, the path ahead is filled with difficulties. Notably, the United States has officially opposed the agreement's text, highlighting the ongoing political divide over the fossil fuel phase-out. Furthermore, several countries are aggressively strengthening their fossil fuel infrastructure to fulfill immediate energy needs, highlighting the conflict between long-term climate ambitions and short-term economic and energy security concerns. Thus, while COP28 produced a symbolic triumph for climate action, the practical implementation of this change will require continuous and concentrated efforts to overcome these major impediments.

A Call for Climate Finance and Global Financial Reforms:
The Africa Climate Summit, held in the run-up to COP28, emphasized the critical need for increased climate finance and radical changes to the global financial system (Colenbrander et al., 2018; Winkler et al., 2024a). African countries are disproportionately affected by climate change, despite having the lowest greenhouse gas emissions (Winkler et al., 2024b). To adapt to and mitigate the effects of climate change, these countries will need significant financial resources. The call for global financial reforms aims to make these resources available to vulnerable countries while also improving the existing financial systems' ability to support sustainable development.

4.5.3 The COP29 Outcomes and Expectations

The COP29 meeting at Baku in 2024 highlighted the crucial necessity of climate finance in solving the world's rising climate problem, with more states participating in negotiations than in prior rounds. However, the meeting was marred by heated disagreement, particularly over the proposed financial pledges from affluent countries to assist developing ones. Richer nations submitted a "totally unacceptable and inadequate" offer of $250 billion in yearly climate finance by 2035, which sparked the debate. This amount, intended to replace the long-standing $100 billion aim, drew strong criticism from developing countries, small island governments, and even academic economists, who saw it as well below the sum required to adequately tackle climate change.

The negotiation process, which lasted longer than expected, showed the wide difference between vulnerable nations' financial requirements and wealthy countries' planned contributions. Developing countries, represented by organizations such as the G77 and the Alliance of Small Island States, underlined the importance of dramatically increasing financial flows, with some pushing for at least $500 billion per year. The dramatic discrepancy between the projected $250 billion and the anticipated $1.3 trillion yearly cost for developing nations to combat climate change highlighted the need for a more robust and equitable financing framework. The African Group of Negotiators' harsh criticism, combined with the acceptance of higher figures by economic specialists such as Lord Nicholas Stern, emphasized the inadequacies of the initial proposal.

The draft declaration from the COP29 chair listed a variety of financial sources, including public and corporate investments, as well as bilateral and international contributions. While international development banks are committed to mobilizing $185 billion by 2030, the entire financial aim remained a source of controversy. The argument focused on wealthy nations' need to take the lead in providing significant financial assistance, reflecting the notion of shared but differentiated duties and capabilities. The push for a more robust financial commitment, in line with economic experts' recommendations and the urgent needs of vulnerable countries, will remain a critical component of future climate negotiations as the world strives to make meaningful progress in mitigating and adapting to the effects of climate change.

4.6 CONCLUSION

Climate financing is a critical component of Africa's and developing countries climate action. Hence, the struggle against climate change is defined by both tremendous accomplishments and ongoing problems. While the continent is actively pursuing climate mitigation and adaptation solutions, the effectiveness of these efforts is dependent on access to adequate and consistent funding. International help is critical, but the constraints of obtaining these money have created a huge deficit. Initiatives such as the Climate and Energy Response Facility (CERF) are critical in addressing this capacity gap by enabling African countries to navigate the complex climate finance landscape and prepare viable proposals.

Key challenges remain, including the need to close the climate finance gap between developed and developing countries, ensure transparency and accountability in fund allocation, and strengthen African nations' technical capacity to effectively manage and implement climate projects. The results of COP28 and the current debates leading up to COP29 highlight the importance of these problems. While progress has been achieved, such as the creation of the Loss and Damage Fund and commitments to clean cooking solutions, financial pledges continue to fall short of the massive investments required to alleviate Africa's susceptibility to climate change. Moving forward, a coordinated effort is needed to expand international cooperation, improve access to climate finance, and guarantee that African countries have the resources and expertise to construct a sustainable and resilient future.

REFERENCES

Arezki, R. (2021). Climate finance for Africa requires overcoming bottlenecks in domestic capacity. *Nature Climate Change*, *11*(11), 888–888.

Baskaran, G., Ekeruche, A., Heitzig, C., Ordu, A. U., & Senbet, L. W. (2023). Financing climate-resilient infrastructure in Africa.

Bayat-Renoux, F., & Glemarec, Y. (2014). *Financing recovery for resilience*. United Nations Development Programme.

Belianska, A., Bohme, N., Cai, K., Diallo, Y., Jain, S., Melina, M. G., ... Zerbo, S. (2022). Climate change and select financial instruments: An overview of opportunities and challenges for sub-Saharan Africa.

Bowman, M., & Minas, S. (2019). Resilience through interlinkage: The green climate fund and climate finance governance. *Climate Policy*, *19*(3), 342–353.

Briera, T., & Lefèvre, J. (2024). Reducing the cost of capital through international climate finance to accelerate the renewable energy transition in developing countries. *Energy Policy*, *188*, 114104.

Chang, A. Y., Cowling, K., Micah, A. E., Chapin, A., Chen, C. S., Ikilezi, G., ... Younker, T. (2019). Past, present, and future of global health financing: A review of development assistance, government, out-of-pocket, and other private spending on health for 195 countries, 1995–2050. *The Lancet*, *393*(10187), 2233–2260.

Chaudhury, A. (2020). Role of intermediaries in shaping climate finance in developing countries—lessons from the green climate fund. *Sustainability*, *12*(14), 5507.

Chirambo, D. (2014). The climate finance and energy investment dilemma in Africa: Lacking amidst plenty. *Journal of Developing Societies*, *30*(4), 415–440.

Colenbrander, S., Dodman, D., & Mitlin, D. (2018). Using climate finance to advance climate justice: The politics and practice of channelling resources to the local level. *Climate Policy*, *18*(7), 902–915.

Colenbrander, S., Pettinotti, L., Cao, Y., Robertson, M., Hedger, M., & Gonzalez, L. (2024). The new collective quantified goal and its sources of funding.

CPI. (2024). Global landscape of climate finance 2024: Insights for COP 29. Available online: climatepolicyinitiative.org/publication/global-landscape-of-climate-finance-2024

Digitemie, W. N., & Ekemezie, I. O. (2024). Assessing the role of climate finance in supporting developing nations: A comprehensive review. *Finance & Accounting Research Journal*, 6(3), 408–420.

Doku, I., Ncwadi, R., & Phiri, A. (2021a). Determinants of climate finance: Analysis of recipient characteristics in sub-Sahara Africa. *Cogent Economics & Finance*, 9(1), 1964212.

Doku, I., Ncwadi, R., & Phiri, A. (2021b). Examining the role of climate finance in the environmental Kuznets curve for sub-Sahara African countries. *Cogent Economics & Finance*, 9(1), 1965357.

Gordon, N. J. (2023). Climate finance: An overview. *Environment: Science and Policy for Sustainable Development*, 65(4), 18–26.

Khan, M., Mfitumukiza, D., & Huq, S. (2021). Capacity building for implementation of nationally determined contributions under the Paris Agreement. In *Making climate action more effective* (pp. 95–106). Routledge.

Kissinger, G., Gupta, A., Mulder, I., & Unterstell, N. (2019). Climate financing needs in the land sector under the Paris Agreement: An assessment of developing country perspectives. *Land Use Policy*, 83, 256–269.

Longhurst, D., & Slater, R. (2022). Financing in fragile and conflict contexts: Evidence, opportunities, and barriers.

Mekonnen, A. (2014). Economic costs of climate change and climate finance with a focus on Africa. *Journal of African Economies*, 23(suppl_2), ii50–ii82.

Mungai, E. M., Ndiritu, S. W., & Da Silva, I. (2021). Unlocking climate finance potential for climate adaptation: Case of climate smart agricultural financing in Sub Saharan Africa. In *African handbook of climate change adaptation* (pp. 2063–2083). Springer.

Mungai, E. M., Ndiritu, S. W., & Da Silva, I. (2022). Unlocking climate finance potential and policy barriers – A case of renewable energy and energy efficiency in sub-Saharan Africa. *Resources, Environment and Sustainability*, 7, 100043.

Nakhooda, S., Caravani, A., Bird, N., Schalatek, L., & America, H. (2011). *Climate finance in sub-Saharan Africa*. Climate Finance Policy Briefs, Heinrich Böll Stiftung North America, Washington, DC, USA and Overseas Development Institute (ODI).

Odhong', C., Wilkes, A., van Dijk, S., Vorlaufer, M., Ndonga, S., Sing'ora, B., & Kenyanito, L. (2019). Financing large-scale mitigation by smallholder farmers: What roles for public climate finance? *Frontiers in Sustainable Food Systems*, 3, 3.

Robertson, M. (2024). The new collective quantified goal on climate finance and its access features.

Rumble, O., & First, J. (2021). Accelerating private sector climate finance in Africa.

Savvidou, G., Atteridge, A., Omari-Motsumi, K., & Trisos, C. H. (2021). Quantifying international public finance for climate change adaptation in Africa. *Climate Policy*, 21(8), 1020–1036.

Scott, A., Worrall, L., & Patel, S. (2018). Aligning energy development and climate objectives in nationally determined contributions. Retrieved on January, 17, 2019.

Stender, F., Moslener, U., & Pauw, W. P. (2020). More than money: Does climate finance support capacity building? *Applied Economics Letters*, *27*(15), 1247–1251.

Strzepek, K. M., & McCluskey, A. (2007). *The impacts of climate change on regional water resources and agriculture in Africa* (Vol. 4290). World Bank Publications.

Swithern, S. (2021). *Accountability in disaster risk financing*. Centre for Disaster Protection (accessed 7 February 2022).

Tamasiga, P., Molala, M., Bakwena, M., Nkoutchou, H., & Onyeaka, H. (2023). Is Africa left behind in the global climate finance architecture: Redefining climate vulnerability and revamping the climate finance landscape – A comprehensive review. *Sustainability*, *15*(17), 13036.

Weikmans, R., van Asselt, H., & Roberts, J. T. (2021). Transparency requirements under the Paris Agreement and their (un) likely impact on strengthening the ambition of nationally determined contributions (NDCs). In *Making climate action more effective* (pp. 107–122). Routledge.

Weischer, L., Warland, L., Eckstein, D., Hoch, S., Michaelowa, A., Koehler, M., … Germanwatch, B. (2016). *Investing in ambition: Analysis of the financial aspects in (intended) nationally determined contributions*. THINK TANK & RESEARCH.

Winkler, H., Watson, C., & Bhandari, P. (2024a). Connecting global stocktake outcomes and COP28 workstreams.

Winkler, H., Watson, C., & Bhandari, P. (2024b). Connecting GST outcomes and COP28 workstreams.

Climate Financing Case Studies

5

5.1 INTRODUCTION

The strengths and shortcomings of climate change mitigation and adaptation are especially visible in Africa and developing countries, which are highly vulnerable to extreme weather variability (Betts et al., 2018; Bedeke, 2023; Williams et al., 2021). While there is a rising acknowledgment of the importance of adaptation, its success is typically hampered by a reactive approach that prioritizes catastrophe recovery above proactive capacity building (Filho et al., 2023; Galaitsi et al., 2024; Williams et al., 2021). This creates a debt cycle and reduces long-term resilience. Addressing the disproportionate economic risks faced by these regions necessitates specialized actions and a shift in investment priorities from recovery to capacity enhancement (Filho et al., 2023; Rodríguez et al., 2007).

Furthermore, the intricacies of climate financing encompass a variety of technologies and methods, which are critical for allocating funds toward climate action. Multilateral climate projects, which use advanced data analysis and remote sensing, are critical for analyzing and monitoring project impacts (Attoh et al., 2024; de Sherbinin & Giri, 2001). However, effective climate finance must prioritize national capacity building and encourage innovative solutions, ensuring that investments yield tangible and long-term results, particularly in vulnerable countries such as Africa and South Asia.

DOI:10.1201/9781003644378-5

5.2 THE STRENGTHS AND WEAKNESSES OF CLIMATE CHANGE MITIGATION AND ADAPTATION

The strengths and limitations of climate change mitigation and adaptation are inextricably linked to a country's competence and responsiveness, as seen most clearly in Africa and emerging countries (Michaelowa et al., 2021b; Shackleton et al., 2015). These areas are particularly sensitive to the extremes of regular climatic variability, and climate change is expected to increase the frequency and severity of extreme weather events (Clarke et al., 2022; Seneviratne et al., 2021). A significant strength is the realization that while adaptation is necessary, its efficiency is dependent on existing adaptive ability and development paradigms (Baard et al., 2014; Clarke et al., 2022; Galaitsi et al., 2024). However, one important drawback is that expenditures tend to focus on disaster recovery over proactive capacity growth (Knemeyer et al., 2009; Rodríguez et al., 2007). This reactive approach frequently results in a debt spiral, with governments borrowing to rebuild following each calamity rather than investing in long-term resilience (Clarke et al., 2022). Improving the ability to manage extreme weather events is critical for limiting economic, social, and human damage; nevertheless, this necessitates a change toward including vulnerability, disaster management, and adaptation into sustainable development planning (Balogun et al., 2020; Conway & Schipper, 2011; Schipper & Pelling, 2006).

Furthermore, the vulnerability of Africa and developing countries to climate change's negative consequences is a major worry in climate policy debates (Brown et al., 2007; Hoegh Guldberg et al., 2018). Despite low temperature increases, studies consistently show a disproportionately high economic risk in certain regions (Prober et al., 2019). This emphasizes the critical necessity for tailored treatments and assistance. While there are different models for assessing vulnerability and adaptation, each has advantages and disadvantages that require constant improvement and adaption to local circumstances (Kelly & Adger, 2000). Lending agencies and donors must change their investment policies to prioritize capacity building over recovery operations and infrastructure construction (Galaitsi et al., 2024; Prober et al., 2019). This move is critical for empowering these countries to address climate change issues constructively.

The strengths and limitations differ in different places of Africa and developing countries (Arezki, 2021; Belianska et al., 2022). For instance, West Africa, despite possessing minimal global GHG emissions, is extremely susceptible due to drought, desertification, and reliance on subsistence agriculture

TABLE 5.1   Strength and weakness of climate change adaptation and mitigation actions

CATEGORY	STRENGTH	WEAKNESS
General Adaptation & Mitigation	• Recognize the necessity for adaptation. • Understanding that effective adaptation is contingent on current adaptive capabilities and development paradigms. • Recognizing that global mitigation necessitates partnership among developed and developing countries.	• The tendency to prioritize disaster recovery over proactive capacity growth. • Debt spirals are the result of reactive measures. • Political challenges, funding inequities, and ensuring equitable involvement in global activities.
Developing Nations (Especially Africa & South Asia)	• An increased awareness of vulnerability to the effects of climate change. • Recognizing the need for customized solutions and help. • Recognizing the need for financial help from wealthy countries.	• Limited resources and capacity for implementing effective adaption methods. • Extreme weather occurrences pose a significant risk. • Despite the fact that temperatures are rising slowly, the economic danger is disproportionally huge. • Insufficient governance processes. • Dependence on subsistence agriculture. • Exposure to vector-borne illnesses. • Agricultural productivity losses.
Financial & Investment	• Recognition of the importance of financial support from developed to underdeveloped countries.	• Investment programs frequently prioritize recovery above capacity building. • Financial discrepancies.
Global Mitigation	• Recognizing that inspiring developing countries requires financial help from wealthier countries to encourage low-carbon growth pathways. • Mechanisms for technology transfer and capacity building are regarded as vital.	• Difficulties in ensuring equitable participation.

(Martín Casas & Remalia Sanago, 2022). A strength is the increased acknowledgment of this vulnerability, but weaknesses remain in the form of inadequate resources and the ability to adopt effective adaptation solutions (Digitemie & Ekemezie, 2024; Savvidou et al., 2021; Williams et al., 2021). Similarly, South Africa, despite having relatively strong adaptive resources, poses major dangers to human health, particularly from vector-borne diseases (Wright et al., 2021). South Asia, with its high poverty rates, poor human development indices, and inadequate governance processes, is particularly exposed to the effects of climate change (Digitemie & Ekemezie, 2024; Duran-Encalada et al., 2017). Studies demonstrating the economic repercussions of climate change, such as agricultural production losses, emphasize the critical need for comprehensive adaptation measures (Galaitsi et al., 2024; Savvidou et al., 2021). Finally, meeting global climate change mitigation targets requires the collaboration of both developed and developing countries (Digitemie & Ekemezie, 2024). A fundamental strength is the recognition that motivating developing countries necessitates financial assistance from affluent countries to encourage low-carbon growth pathways (Arezki, 2021; Eniolorunda, 2014). Mechanisms for technology transfer and capacity building are critical for allowing these countries to effectively participate to global mitigation initiatives (Digitemie & Ekemezie, 2024). However, problems remain in the form of political hurdles, financing disparities, and the difficulties of assuring fair participation (Table 5.1). Building a climate-resilient future requires addressing these limitations and exploiting the assets of both mitigation and adaptation initiatives (Table 5.1).

5.3 HOW DIFFERENT TECHNOLOGIES USED TO FACILITATE CLIMATE FINANCE

The climate finance environment is complicated and changing, with numerous technologies and procedures playing critical roles in enabling the flow of cash to combat climate change (Brown et al., 2010; Digitemie & Ekemezie, 2024). Multilateral climate programs, like the Global Environment Facility (GEF), Climate Investment Funds (CIFs), Adaptation Fund (AF), and Green Climate Fund (GCF), have played an important role in directing resources to developing countries (Digitemie & Ekemezie, 2024; Reinsberg et al., 2020). These programs use a variety of technologies, including sophisticated data gathering and analysis tools, to evaluate project proposals, track progress, and determine impact (Brown et al., 2010; Reinsberg et al., 2020; Tracy, 2024). Remote sensing and geographic information systems (GISs) are increasingly being used to monitor deforestation, land-use changes, and the effectiveness of adaptive strategies (Eniolorunda, 2014; Karazhanova & Kim, 2024; Prober et al., 2019). Furthermore, financial innovations like online platforms and digital payment systems make it easier to disburse and track funds, increasing transparency and efficiency (Eniolorunda, 2014; Michaelowa et al., 2021b). The recently developed Climate Project Explorer, an AI-powered application, shows how modern technology can improve the transparency and accessibility of climate project information by allowing users to explore projects sponsored by the major Multilateral Climate Funds (Besinga & Mukete, 2025; Karazhanova & Kim, 2024; Reinsberg et al., 2020). These tools improve the ability to monitor and assess the performance of climate finance, providing important evidence to governments and stakeholders about the impact of their investments.

The efficiency of climate funding in making a tangible difference is a significant issue (Michaelowa et al., 2021b; Prober et al., 2019). Governments in contributing countries want credible evidence that their investments are producing positive results in order to justify sustained or increasing commitments (Besinga & Mukete, 2025; Karazhanova & Kim, 2024). According to the reports, climate funds have broken new ground by aiding countries in tackling the implications of climate change for development (Besinga & Mukete, 2025; Brown et al., 2010). Reduction financing is carefully directed to developing nations with high and rising GHG emissions, maximizing prospects for effective reduction (Hoegh Guldberg et al., 2018). Adaptation funding, on the other hand, focuses on the poorest and most vulnerable countries, particularly in sub-Saharan Africa and South Asia, where the effects of climate change are most severe (Besinga & Mukete, 2025). Success stories, such as the renewable energy expansion in Mexico and the reforestation initiatives in Brazil's Amazon Fund, highlight the funds' ability to promote substantial change. The AF's emphasis on locally led adaptation (LLA) through direct and expanded direct access, which enables vulnerable communities to execute their own climate resilience measures, has been shown to be particularly beneficial (Karazhanova & Kim, 2024; Michaelowa et al., 2021b; Williams et al., 2021). The recent approvals of new projects by the AF, including a large amount dedicated to direct access and LLA, demonstrate this commitment (Besinga & Mukete, 2025; Brown et al., 2010). The fund's commitment to boosting public investment, as indicated by CMA 6 decisions to quadruple annual outflows by 2030, demonstrates a desire to expand its effect (Besinga & Mukete, 2025; Eniolorunda, 2014).

However, challenges persist, and climate funding must become more adaptable and risk-tolerant, fostering innovation and promoting the use of innovative technologies (Karazhanova & Kim, 2024). Transparency in reporting results and impact is also important, as are attempts to reduce transaction costs

TABLE 5.2   The key climate finance agreements and events timeline

YEAR	EVENT AND AGREEMENT	DETAILED DESCRIPTIONS
1992	• UN Framework Convention on Climate Change (UNFCCC)	• Creates a framework for global action to reduce GHG emissions. Recognizes industrialized countries' historical responsibilities and commits them to mobilizing climate financing for underdeveloped nations.
2008	• Establishing Climate Investment Funds (CIFs)	• Established as global funds managed by the World Bank to test new approaches to climate finance at scale.
2009	• The Copenhagen Accord	• Developed countries have pledged to raise $100 billion annually from public and private sources by 2020.
2010	• Agreements signed in Cancun	• Reaffirms the $100 billion commitment every year and acknowledges the necessity for balanced adaptation financing.
2010–2012	• Pledges from Developed Countries	• Commitment to delivering funding worth about $30 billion.
2014 (November)	• Green Climate Fund (GCF) Commitments	• The GCF raises more than $9 billion during its resource mobilization operation.
2015	• COP21	• Paris Agreement
2019	• Paris climate accord	• The Adaptation Fund (AF) serves the Paris Agreement.
2022	• COP27	• Adaptation Fund (AF) Payouts
2023–2027	• Medium-Term Strategy for the Adaptation Fund (AF)	• The emphasis is on locally led adaptation, scaling programs, and reinforcing key pillars.
2024	• The Adaptation Fund Board Meeting	• Approves $82 million for 12 new initiatives, creates locally led adaptation (LLA) financing windows, and sets a $300 million resource mobilization target.
2024 (November)	• COP29	• Climate Project Explorer Launch

(continued)

TABLE 5.2   (Continued)

YEAR	EVENT AND AGREEMENT	DETAILED DESCRIPTIONS
2024 (November)	• COP29)	• Adaptation Fund (AF) Pledges
2024 (COP29)	• CMA 6 Decisions	• To strengthen climate finance, efforts are underway to significantly increase annual funding outflows from key financial mechanisms by 2030, as well as ambitious goals to increase overall climate finance to developing countries to $300 billion by 2030 and $1.3 trillion by 2035, while also establishing a carbon market mechanism that directs a portion of its proceeds to the Adaptation Fund.
2025 (Feb 3)	• Cooperation between the Adaptation Fund (AF) and the Fund for Responding to Loss and Damage (FRLD)	• Framework for collaboration to identify synergies and prevent duplication.
2030	• CMA 6 Target	• Aim to treble yearly climate fund outflows from 2022 levels.
2035	• NCQG Target	• Triple climate finance to developing countries from $100 billion to $300 billion each year. Aim to increase climate finance to developing nations to US$ 1.3 trillion annually.

and streamline decision-making processes. Furthermore, money should be allocated to a larger range of stakeholders, such as governments, enterprises, and communities, with a focus on national capacity building (Besinga & Mukete, 2025; Hoegh Guldberg et al., 2018). Engaging the private sector and creating new collaborations with financial institutions involved in climate-related sectors, notably infrastructure, is critical for mobilizing larger amounts of financing (Karazhanova & Kim, 2024; Michaelowa et al., 2021b; Salimi Turkamani, 2024). The need to streamline and unify the climate financing infrastructure is also evident, as the proliferation of funds causes overlap and inefficiencies (Table 5.2). The GCF, with its vast resources and worldwide reach, has the ability to address these issues and play a critical role in scaling up climate finance (Hoegh Guldberg et al., 2018). The AF's collaboration with other funds, such as the Fund for Responding to Loss and Damage, to find synergies and avoid duplication, combined with its focus on increasing access to adaptation finance through LLA programs, demonstrates a proactive approach to improving climate finance effectiveness (Salimi Turkamani, 2024). The recognition of the AF's accomplishments and role in assisting developing nations, as noted in the CMP 19 and CMA 6 rulings, emphasizes its significance in the global climate financing architecture (Table 5.2).

5.4 GENDER AND MARGINALIZED GROUPS' CONTRIBUTION TO CLIMATE FINANCE

Gender concerns and the inclusion of marginalized populations in climate financing frameworks are not only ethical imperatives, but also key components of effective climate action (Sultana, 2022; Terry, 2009). Cultural barriers and existing inequities, compounded by climate-related threats such as floods, disease outbreaks, and water scarcity, have a disproportionate impact on women's livelihoods (Sarma et al., 2024; Terry, 2009). This fact emphasizes the importance of understanding women not only as vulnerable populations, but also as powerful change agents (Table 5.3). Analyses of Nationally Determined Contributions (NDCs) across African states show that gender is widely acknowledged, with over 85% of acts citing it (Michaelowa et al., 2021a; Sultana, 2022; Weikmans et al., 2021). However, the regional distribution of these acts varies greatly, underscoring the importance of targeted interventions (Prober et al., 2019; Sarma et al., 2024). Women provide invaluable coping strategies, environmental knowledge, and ecosystem services, creating untapped prospects for climate resilience (Terry, 2009). Empowering women with access to knowledge, education, training, and secure land rights is critical for minimizing climate risks and realizing their potential as climate leaders (Table 5.3).

Despite rising understanding of the gender-climate nexus, there remains a large gap between high-level policy goals and on-the-ground reality (Prober et al., 2019). Climate change definitely exacerbates existing gender disparities, impeding progress toward global goals including the Paris Agreement, the UN 2030 Sustainable Development Goals, and the African Union's Agenda 2063 (Table 5.3). Gender and climate finance dynamics have a substantial impact on economic and political environments, in addition to environmental concerns (Dwivedi et al., 2022; Terry, 2009). A more in-depth examination of

TABLE 5.3 The gender and marginalized groups' contributions to climate finance

CATEGORY	CONTRIBUTION/ROLE	CHALLENGES/BARRIERS	STRATEGIES FOR IMPROVEMENT
Women	• Local knowledge of ecosystems and resource management. • Effective coping strategies during climate-related disasters. • Leadership in community-based adaptation initiatives. • Participation in sustainable agriculture.	• Limited access to financial resources and credit. • Lack of secure land rights. • Exclusion from decision-making processes. • Cultural and social norms that restrict their participation.	• Provide targeted funding and microfinance programs. • Strengthen women's land rights. • Ensure women's representation in climate finance governance. • Invest in education and capacity building.
Indigenous Communities	• Traditional ecological knowledge and sustainable resource management practices. • Guardianship of biodiversity-rich areas. • Resilience in the face of environmental change.	• Lack of recognition of traditional land rights. • Exclusion from mainstream climate finance mechanisms. • Loss of traditional knowledge due to cultural erosion. • Lack of participation in decision-making.	• Recognize and respect indigenous land rights. • Establish direct funding mechanisms for indigenous-led climate initiatives. • Integrate traditional knowledge into climate adaptation strategies. • Ensure Free Prior and Informed Consent.
People with Disabilities	• Unique perspectives on accessibility and resilience. • Advocacy for inclusive climate action. • Knowledge of accessible adaptation strategies.	• Exclusion from climate finance planning and implementation. • Lack of accessible infrastructure and information. • Discrimination and stigma.	• Ensure accessibility in climate finance projects. • Include people with disabilities in decision-making bodies. • Provide targeted support for disability-inclusive adaptation. • Collect disaggregated data.

| **Economically Marginalized Groups** | • Experience of climate impacts on livelihoods and food security.
• Knowledge of coping mechanisms in resource-scarce environments.
• Participation in community-based disaster risk reduction. | • Limited access to financial resources and social safety nets.
• Vulnerability to climate-induced displacement.
• Lack of access to information and technology. | • Provide social protection programs and micro insurance.
• Invest in climate-resilient livelihoods.
• Improve access to information and early warning systems.
• Community-based adaptation project |

these interactions is required for the development of successful climate change and poverty-eradication programs based on gender equality and women's empowerment. Climate finance programs must actively confront and transform current gender disparities in mitigation, adaptation, and disaster management measures (Dwivedi et al., 2022).

The three critical issues must be considered for achieving genuine gender equity in climate funding, notably in programs such as tree planting, smart agriculture, and catastrophe information dissemination. To begin, it is critical to understand how men and women's differing incentives and preferences are molded by their lived experiences and priorities, as well as the variables that enable or limit their access to resources. Second, both official and informal participatory forums must provide meaningful opportunities for women and men to make decisions that empower them. Finally, it is critical to satisfy women's practical and strategic demands while resolving any underlying inconsistencies (Dwivedi et al., 2022; Terry, 2009). Fundamentally, the participation of gender and disadvantaged groups in climate financing necessitates a commitment to questioning and reforming unfair socio-political arrangements, such as inequitable land rights, that perpetuate gender imbalances (Sultana, 2022). This comprehensive strategy ensures that climate finance not only reduces climate consequences, but also promotes a more equal and just society.

5.5 CLIMATE FINANCE MECHANISMS FOCUSING IN AFRICA AND DEVELOPING COUNTRIES

Climate finance arrangements suited to Africa and developing countries are critical in the global fight against climate change (Bedeke, 2023; Sultana, 2022). Since the inception of climate efforts, the primary international focus has been on developing a global framework to reduce greenhouse gas emissions from developed nations, who have historically and continue to be the largest contributors (Lamb et al., 2021). While reducing emissions in affluent countries is critical, the focus is gradually turning to developing and implementing effective policies for climate-friendly measures in developing countries (Bedeke, 2023; Hoegh Guldberg et al., 2018; Sarma et al., 2024). Though studies of these difficulties might be applied to wealthy countries, the primary focus here is on the integration of climate and development goals in the developing world.

Although developing countries' per capita greenhouse gas emissions are expected to stay much lower than those of industrialized nations for the foreseeable future, their total emissions are fast increasing (Lamb et al., 2021; Prober et al., 2019). Over the next decade or two, their total emissions are expected to exceed those of industrialized countries. As a result, even with substantial mitigation efforts in industrialized countries, the most likely emission scenarios indicate that atmospheric greenhouse gas concentrations cannot be stabilized until emerging countries drastically reduce their emissions below business-as-usual estimates (Digitemie & Ekemezie, 2024). As a result, future climate initiatives must explicitly recognize and meet developing countries' essential development needs. This recognition is critical for productively engaging these nations in their shared responsibility for climate protection. Climate finance must effectively aid the transition to low-carbon, climate-resilient development options that promote economic growth while also reducing poverty. To enable developing countries to move beyond old development paradigms and adopt sustainable alternatives, significant financial and technology aid, capacity building, and policy support are required. Without such targeted and comprehensive support, the global effort to prevent climate change will be seriously jeopardized.

5.6 OVERVIEW AND IMPACTS OF CLIMATE FINANCE PROJECTS IN AFRICA

Climate financing programs in Africa and other developing nations are critical for negotiating the complicated interactions of commerce, development, and climate change (Arezki, 2021; Lamb et al., 2021). Technical assistance focused on upgrading trade policy is critical because it enables these nations to develop and implement plans that increase their competitiveness in changing global markets (Bedeke, 2023; Dwivedi et al., 2022). Climate change is changing the face of international trade, altering manufacturing patterns, trade restrictions, and the standards to which traded goods must adhere. As a result, developing countries require strong help to adapt to these changes and capitalize on new opportunities.

The impact of climate change on African agriculture is especially concerning (Arezki, 2021; Lamb et al., 2021). The scarcity of country-specific studies, owing largely to data limitations, impedes a thorough understanding of the difficulties (Bedeke, 2023). Existing research frequently focuses on a small range of crops, particularly grains, while ignoring the larger spectrum

TABLE 5.4 The overview and impacts of climate finance projects in Africa

CATEGORY	KEY ISSUES/CHALLENGES	IMPACTS	CLIMATE FINANCE PROJECT FOCUS/NEEDS
Trade & Policy	• Adapting trade policies to climate change. • Navigating changing global market demands. • Meeting new trade standards & regulations.	• Increased or decreased competitiveness. • Changes in export revenue. • Disruptions in supply chains.	• Technical assistance for trade policy development. • Capacity building for compliance with new standards. • Support for diversification of export markets.
Agriculture	• Climate-induced crop yield reductions. • Data scarcity for country-specific analysis. • Over-reliance on limited crop types. • Vulnerability of agriculture-dependent economies.	• Reduced agricultural output and export revenues. • Food security risks. • Economic instability. • Potential loss of over 50% of agricultural output in some nations by 2080. • 20% drop in Malawi agricultural export revenues.	• Investment in climate-resilient agriculture. • Data collection and research for targeted interventions. • Diversification of crops and farming practices. • Implementation of both top-down and bottom-up adaptation strategies.
Low-Carbon Transition	• Shifting to sustainable, low-carbon economies. • Capitalizing on new market opportunities. • Unlocking developing countries' abatement potential.	• Creation of new green industries and jobs. • Reduced greenhouse gas emissions. • Enhanced energy security. • Global greenhouse gas abatement options are located 70% in developing nations. • 90% of terrestrial carbon opportunities are located in developing nations, accounting for 30% of global GHG abatement.	• Investment in renewable energy and clean technologies. • Support for sustainable land management and forestry. • Carbon finance mechanisms. • Targeted funding to unlock GHG abatement potential.

of agricultural outputs (Dwivedi et al., 2022). This error is crucial because climate-related declines in agricultural output will have far-reaching economic effects. Even accounting for the possibly favorable impacts of carbon fertilization, projections show that some agriculture-dependent nations could lose more than half of their entire agricultural output by 2080. In Malawi, this amounts to a potential 20% drop in agricultural export revenues, emphasizing the critical need for adaptation solutions (Table 5.4). Aside from just "climate-proofing" existing agricultural methods, which necessitates both top-down and bottom-up techniques, adaptation must also include a shift to a low-carbon global economy. This change offers both obstacles and possibilities (Lamb et al., 2021; Sultana, 2022). The expanding global demand for low-carbon products and services opens up new markets for emerging countries (Bedeke, 2023; Lamb et al., 2021). Crucially, nearly 70% of global greenhouse gas abatement options are located in the developing countries. Furthermore, these regions contain 90% of terrestrial carbon opportunities, accounting for 30% of global GHG abatement potential (Table 5.4). This emphasizes developing countries' important role in climate change mitigation and the importance of targeted climate funding to help them transition to sustainable, low-carbon economies.

5.7 CONCLUSION

The case studies highlight the crucial need for a paradigm shift in climate change mitigation and adaptation measures, especially in Africa and other developing countries (Prober et al., 2019; Sultana, 2022). These regions, which are disproportionately affected by climate change due to their reliance on climate-sensitive sectors and inadequate adaptation capacity, demand a shift from reactive disaster recovery to proactive capacity building. This transformation involves climate finance instruments that are customized to their specific development needs, promoting the transition to low-carbon, climate-resilient paths. Furthermore, the efficiency of climate funding is dependent on the strategic use of new technology, such as remote sensing and GIS, for project monitoring and impact assessment (Eniolorunda, 2014; Sarma et al., 2024). As a result, to ensure that investments produce tangible and long-term results, multilateral climate funds must promote transparency, lower transaction costs, and increase stakeholder involvement. Gender considerations and the inclusion of vulnerable groups are especially critical, as equitable climate action is dependent on resolving existing disparities. Empowering women and guaranteeing their involvement in decision-making processes is critical for using their knowledge and capacities to develop climate resilience (Sultana, 2022; Terry, 2009). Finally, developing countries have significant greenhouse gas abatement potential, emphasizing the importance of targeted climate finance in unlocking this potential and facilitating their transition to a sustainable future. This necessitates not just financial and technological aid, but also capacity building and policy support to enable these countries to pursue low-carbon development strategies. Recognizing these characteristics is critical for building a climate-resilient future that takes into account Africa's particular vulnerabilities and opportunities.

REFERENCES

Arezki, R. (2021). Climate finance for Africa requires overcoming bottlenecks in domestic capacity. *Nature Climate Change*, *11*(11), 888–888.

Attoh, E. M., Amarnath, G., Sahana, V., Okem, A., Panjwani, S., Shashanka, G., ... Alahacoon, N. (2024). Development of climate smart governance dashboard to support national climate action and resilience building.

Baard, S. K., Rench, T. A., & Kozlowski, S. W. (2014). Performance adaptation: A theoretical integration and review. *Journal of Management*, *40*(1), 48–99.

Balogun, A.-L., Marks, D., Sharma, R., Shekhar, H., Balmes, C., Maheng, D., ... Salehi, P. (2020). Assessing the potentials of digitalization as a tool for climate change adaptation and sustainable development in urban centres. *Sustainable Cities and Society*, *53*, 101888.

Bedeke, S. B. (2023). Climate change vulnerability and adaptation of crop producers in sub-Saharan Africa: A review on concepts, approaches and methods. *Environment, Development and Sustainability*, *25*(2), 1017–1051.

Belianska, A., Bohme, N., Cai, K., Diallo, Y., Jain, S., Melina, M. G., ... Zerbo, S. (2022). Climate change and select financial instruments: An overview of opportunities and challenges for sub-Saharan Africa.

Besinga, L. E., & Mukete, T. N. M. (2025). Technological innovations for climate adaptation and peacebuilding: A holistic approach to resource conflict and environmental challenges. *Open Journal of Applied Sciences*, *15*(1), 285–304.

Betts, R. A., Alfieri, L., Bradshaw, C., Caesar, J., Feyen, L., Friedlingstein, P., ... Morfopoulos, C. (2018). Changes in climate extremes, fresh water availability and vulnerability to food insecurity projected at 1.5 C and 2 C global warming with a higher-resolution global climate model. *Philosophical Transactions of the Royal Society A: Mathematical, Physical and Engineering Sciences*, *376*(2119), 20160452.

Brown, J., Cantore, N., & te Velde, D. W. (2010). *Climate financing and development*. Overseas Development Institute.

Brown, O., Hammill, A., & McLeman, R. (2007). Climate change as the 'new' security threat: Implications for Africa. *International Affairs*, *83*(6), 1141–1154.

Clarke, B., Otto, F., Stuart-Smith, R., & Harrington, L. (2022). Extreme weather impacts of climate change: An attribution perspective. *Environmental Research: Climate*, *1*(1), 012001.

Conway, D., & Schipper, E. L. F. (2011). Adaptation to climate change in Africa: Challenges and opportunities identified from Ethiopia. *Global Environmental Change*, *21*(1), 227–237.

de Sherbinin, A., & Giri, C. (2001). Remote sensing in support of multilateral environmental agreements: What have we learned from pilot applications? Paper presented at the open meeting of the human dimensions of global environmental change research community. Rio de Janeiro.

Digitemie, W. N., & Ekemezie, I. O. (2024). Assessing the role of climate finance in supporting developing nations: A comprehensive review. *Finance & Accounting Research Journal*, *6*(3), 408–420.

Duran-Encalada, J. A., Paucar-Caceres, A., Bandala, E., & Wright, G. (2017). The impact of global climate change on water quantity and quality: A system dynamics approach to the US–Mexican transborder region. *European Journal of Operational Research*, *256*, 567–581.

Dwivedi, Y. K., Hughes, L., Kar, A. K., Baabdullah, A. M., Grover, P., Abbas, R., … Bunker, D. (2022). Climate change and COP26: Are digital technologies and information management part of the problem or the solution? An editorial reflection and call to action. *International Journal of Information Management*, *63*, 102456.

Eniolorunda, N. (2014). Climate change analysis and adaptation: The role of remote sensing (Rs) and geographical information system (Gis). *International Journal of Computational Engineering Research*, *4*(1), 41–51.

Filho, W. L., Minhas, A., Schmook, B., Mardero, S., Sharifi, A., Paz, S., … Skouloudis, A. (2023). Sustainable development goal 13 and switching priorities: Addressing climate change in the context of pandemic recovery efforts. *Environmental Sciences Europe*, *35*(1), 6.

Galaitsi, S. E., Corbin, C., Cox, S. A., Joseph, G., McConney, P., Cashman, A., … Trump, B. D. (2024). Balancing climate resilience and adaptation for Caribbean Small Island Developing States (SIDS): Building institutional capacity. *Integrated Environmental Assessment and Management*, *20*(5), 1237–1255.

Hoegh Guldberg, O., Jacob, D., Taylor, M., Bindi, M., Brown, S., Camilloni, I. A., … Engelbrecht, F. (2018). Impacts of 1.5 C global warming on natural and human systems.

Karazhanova, A., & Kim, T. H. (2024). Leveraging digital technologies and data for climate action.

Kelly, P. M., & Adger, W. N. (2000). Theory and practice in assessing vulnerability to climate change and facilitating adaptation. *Climatic Change*, *47*(4), 325–352.

Knemeyer, A. M., Zinn, W., & Eroglu, C. (2009). Proactive planning for catastrophic events in supply chains. *Journal of Operations Management*, *27*(2), 141–153.

Lamb, W. F., Wiedmann, T., Pongratz, J., Andrew, R., Crippa, M., Olivier, J. G., … House, J. (2021). A review of trends and drivers of greenhouse gas emissions by sector from 1990 to 2018. *Environmental Research Letters*, *16*(7), 073005.

Martín Casas, N., & Remalia Sanogo, A. (2022). Climate finance in West Africa: Assessing the state of climate finance in one of the world's regions worst hit by the climate crisis.

Michaelowa, A., Espelage, A., Lieke't Gilde, N. K., Censkowsky, P., Greiner, S., Ahonen, H.-M., … Dalfiume, S. (2021a). Article 6 readiness in updated and second NDCs. *Perspectives climate group & climate focus*, 1–64.

Michaelowa, A., Hoch, S., Weber, A.-K., Kassaye, R., & Hailu, T. (2021b). Mobilising private climate finance for sustainable energy access and climate change mitigation in sub-Saharan Africa. *Climate Policy, 21*(1), 47–62.

Prober, S. M., Doerr, V. A., Broadhurst, L. M., Williams, K. J., & Dickson, F. (2019). Shifting the conservation paradigm: A synthesis of options for renovating nature under climate change. *Ecological Monographs, 89*(1), e01333.

Reinsberg, B., Shishlov, I., Michaelowa, K., & Michaelowa, A. (2020). Climate change-related trust funds at the multilateral development banks.

Rodríguez, H., Quarantelli, E. L., Dynes, R. R., Smith, G. P., & Wenger, D. (2007). Sustainable disaster recovery: Operationalizing an existing agenda. *Handbook of disaster research* (pp. 234–257). Springer.

Salimi Turkamani, H. (2024). The loss and damage fund: A solution to interpretive conflicts of responsibility for climate change? *Netherlands International Law Review*, 71, 1–26.

Sarma, H. H., Borah, S. K., Dutta, N., Sultana, N., Nath, H., & Das, B. C. (2024). Innovative approaches for climate-resilient farming: Strategies against environmental shifts and climate change. *International Journal of Environment and Climate Change, 14*(9), 217–241.

Savvidou, G., Atteridge, A., Omari-Motsumi, K., & Trisos, C. H. (2021). Quantifying international public finance for climate change adaptation in Africa. *Climate Policy, 21*(8), 1020–1036.

Schipper, L., & Pelling, M. (2006). Disaster risk, climate change and international development: Scope for, and challenges to, integration. *Disasters, 30*(1), 19–38.

Seneviratne, S. I., Zhang, X., Adnan, M., Badi, W., Dereczynski, C., Luca, A. D., … Lewis, S. (2021). Weather and climate extreme events in a changing climate.

Shackleton, S., Ziervogel, G., Sallu, S., Gill, T., & Tschakert, P. (2015). Why is socially-just climate change adaptation in sub-Saharan Africa so challenging? A review of barriers identified from empirical cases. *Wiley Interdisciplinary Reviews: Climate Change, 6*(3), 321–344.

Sultana, F. (2022). Critical climate justice. *The Geographical Journal, 188*(1), 118–124.

Terry, G. (2009). No climate justice without gender justice: An overview of the issues. *Gender & Development, 17*(1), 5–18.

Tracy, S. J. (2024). *Qualitative research methods: Collecting evidence, crafting analysis, communicating impact*. John Wiley & Sons.

Weikmans, R., van Asselt, H., & Roberts, J. T. (2021). Transparency requirements under the Paris Agreement and their (un) likely impact on strengthening the ambition of nationally determined contributions (NDCs). In *Making climate action more effective* (pp. 107–122). Routledge.

Williams, P. A., Simpson, N. P., Totin, E., North, M. A., & Trisos, C. H. (2021). Feasibility assessment of climate change adaptation options across Africa: An evidence-based review. *Environmental Research Letters, 16*(7), 073004.

Wright, C. Y., Kapwata, T., du Preez, D. J., Wernecke, B., Garland, R. M., Nkosi, V., … Norval, M. (2021). Major climate change-induced risks to human health in South Africa. *Environmental Research, 196*, 110973.

Recommendation and Future Perspectives in Effective Climate Financing

6

6.1 RECOMMENDATIONS

Climate Finance Gap in Africa and Emerging Nations (Chapter 1)
To effectively close the climate finance gap in Africa and emerging nations, a multifaceted international effort is required.

- First and foremost, policy frameworks must be improved. This includes expediting clearance processes, establishing clear norms, and incentivizing climate-friendly investments to attract domestic and international actors. De-risking methods, such as guarantees and insurance, should also be developed to promote private sector involvement.
- Second, capacity building is needed. Investments in training programs, technical assistance, and knowledge-sharing activities are essential to provide African institutions with the skills and competence they need to properly handle climate money. This will increase openness, accountability, and efficiency in project implementation, making these countries more appealing for climate investment.

DOI:10.1201/9781003644378-6

- Third, alternative funding sources must be considered. Green bonds and debt-for-nature swaps are two innovative financing techniques that show promise as alternatives to existing routes. Diversifying the resource pool will minimize dependency on traditional aid and hasten the transition to a climate-resilient future.
- Furthermore, increased international cooperation is essential. Developed countries must keep their promises to give financial help by increasing the amount of climate funding allocated to Africa and enhancing access to these funds. To enable a reasonable and fair global climate response, ethical factors such as equitable cost and reward distribution must be addressed.

Finally, national legislation plays an important impact. Ambitious legislation to decrease emissions and strengthen climate resilience should be priority. Strengthening national positions through law enables more ambitious aims in international discussions.

Africa's Vulnerability to Climate Change (Chapter 2)

To properly address Africa's rising climate problem, a fundamental change in climate finance is required. To address the continent's extreme vulnerability, developed countries must significantly increase their financial contributions, with a focus on adaptation projects and the critical Loss and Damage Fund. Furthermore, more equal allocation of this money is required, with least developed countries (LDCs) receiving a larger part. A major shift from loan-based financing to grant-based support for adaptation initiatives is also required to avoid worsening existing high debt burdens in vulnerable countries.

Improving accountability and openness in climate finance is crucial. This demands the development of clear definitions and rigorous monitoring methods for tracking financial flows and ensuring their effective use. Addressing existing uncertainties, coverage constraints, and dependability difficulties in climate finance estimates is critical for proper tracking. Furthermore, increased openness in money disbursement is required to boost the currently low disbursement rates for adaptation initiatives. Equally crucial is the appropriate monitoring of Nationally Determined Contributions (NDCs) to guarantee compliance with declared obligations.

Given the reliance on domestic funding, strengthening national policies and domestic financial sources is critical for shaping climate finance flows. Developing strong local capital markets is critical for facilitating long-term climate investments. Furthermore, a stronger emphasis on adaptation investment, which is now lagging behind mitigation money, is critical to meeting African countries' immediate requirements. Developed countries must also honor their commitments to the Loss and Damage Fund, ensuring that it is appropriately funded to assist countries dealing with climate-related disasters.

Promoting sustainable development and renewable energy transitions are both critical. Clean cooking solutions should receive more support in order to promote public health and reduce deforestation. Africa could use global initiatives such as the global renewables and energy efficiency pledge and the global methane pledge to hasten its transition to sustainable energy. Clear instructions and rules for public-private partnerships (PPPs), forest carbon financing mechanisms, forest concessions, and private sector participation in carbon financing are critical tools for changing the forestry industry.

Finally, addressing the possible negative consequences of huge financial inflows necessitates strong institutional frameworks. Price corridors and sovereign wealth funds should be used to reduce financial risks, reducing volatility, Dutch disease, and rent-seeking. This comprehensive strategy is required to ensure that climate finance effectively contributes to Africa's resilience and sustainable development.

Improving Climate Resilience in Africa (Chapter 3)

To improve global climate resilience, the UAE Framework for Global Climate Resilience (UAE FGCR) should be strengthened with measurable measures of adaptation success, and adaptation efforts should be directed toward the most pressing challenges. Nationally, increasing adaptation spending, particularly for the African health sector, is critical for improving early warning systems, constructing robust infrastructure, and training healthcare professionals. Comprehensive National Adaptation Plans (NAPs) should be prepared and implemented in accordance with Climate-Resilient Green Economy (CRGE) policies, and climate change considerations should be incorporated into all development plans. To properly track progress, quantitative indicators should be utilized to measure adaptation, and a strong framework for monitoring, evaluation, reporting, verification, and certification (MERVC) should be built for forestry-based climate mitigation projects. Furthermore, adaptation programs should prioritize people, empower communities, and address underlying health inequities. Increased investment for mitigation and adaptation, as well as the continuous creation of NDCs focusing on sustainable agriculture, forest protection, renewable energy, and energy efficiency, is critical. Priority should be given to areas that have dual benefits: carbon reduction and economic development.

A large increase in financial resources is required for African healthcare systems to improve resilience, including cash for early warning systems, resilient infrastructure, and professional development. Key ministries must establish technical capabilities for efficient climate finance mobilization, such as project pipeline creation, environmental and social protections, economic and financial analysis, climate budget tagging and tracking, and non-traditional finance access. Detailed project proposals for funding mechanisms such as the Green Climate Fund (GCF) and Loss and Damage Funds should be created, as should comprehensive economic and financial analyses of climate projects. Subnational and federal project management abilities should be improved, and a data management platform based on IT should be deployed. Effective project execution and tracking procedures are required to ensure the effectiveness of climate finance. A uniform definition of "climate finance" and unambiguous tracking methodology should be established, as well as increased data collection and transparency about climate money flows, including NGO-led projects. Green budget tagging (GBT) should be introduced to improve transparency and accountability in climate-related expenditures.

Implementation should prioritize increasing technical competence within government ministries to generate and manage climate money, as well as improving agro-meteorological services. Institutional structures for climate action should be improved, with the ministry of planning and development acting as a key coordination point. Climate change adaptation should be built into development plans, with a "Development Pathways" approach used to examine climate impacts and adaptation choices. For forestry-based climate mitigation programs, rigorous MERVC systems should be created, properly assessing carbon stock changes while also taking into account co-benefits such as biodiversity protection. Finally, GBT should be used to track climate-related spending and increase openness in public and private climate finance programs, allowing for more accurate evaluation of budget components' climate impact.

Improving Climate Financing Capacity in Africa (Chapter 4)

The major recommendation is to dramatically increase capacity-building programs within African countries to improve their access to and use of climate money. This includes providing African governments with the technical knowledge and proposal preparation skills required to negotiate the complex climate funding ecosystem. The Climate and Energy Response Facility (CERF) is emphasized as a possible model for accomplishing this, with a focus on producing strong proposals, managing funds efficiently, and implementing long-term climate mitigation and adaptation initiatives. It is critical to overcome problems such as a lack of proposal expertise, shortcomings in financial analysis, and poor project management.

Fostering strategic alliances and collaboration is another important facet of increasing climate action and resource mobilization. Encourage government-to-government engagement for information sharing and finance access, promote PPPs to generate extra financial resources, and include the business sector in climate-resilient development. These initiatives will be bolstered by more international collaboration and improved negotiation support from African governments.

Furthermore, the manuscript explicitly urges strategic partnerships and increased negotiating support to help Africa realize its potential for meeting its

NDCs and creating climate resilience. This involves encouraging government-to-government collaboration, utilizing international ties to seek funds, and promoting knowledge sharing and collective action. The emphasis is on encouraging African countries to participate actively in climate change mitigation rather than simply receiving money.

Finally, there is a strong appeal for global financial changes and more accessible and fair climate finance. The debates at COP28 and COP29 highlighted the significance of continuous negotiations on the New Collective Quantified Goal (NCQG), as well as addressing concerns such as debt and equitable burden sharing among wealthy countries. The statement underlines the importance of moving beyond mere pledges to tangible results, focusing on concrete figures, verifiable progress, and openness in climate finance accounting. It also supports for increasing public sector contributions and concessionary funding, particularly for fragile countries, while empowering local communities and subnational governments through mechanisms such as developed climate finance (DCF).

Shift from Reactive Disaster Recovery to Proactive Capacity Creation (Chapter 5)

A major shift is required in climate finance, from reactive disaster recovery to proactive capacity creation. This method would increase long-term resilience, particularly in Africa and other emerging countries. Most importantly, climate finance instruments must be adapted to these nations' individual developmental demands, allowing them to transition to low-carbon, climate-resilient pathways. Incorporating modern technology like remote sensing and GIS will improve project monitoring and effect assessment. Furthermore, AI-powered applications can increase the transparency and accessibility of climate project information, resulting in better accountability.

Transparency and stakeholder involvement are critical for successful climate finance. This involves fostering transparent reporting and impact evaluations, enhancing stakeholder involvement, and lowering transaction costs in multilateral climate programs. Gender concerns and marginalized groups' engagement in climate financing plans are critical for ensuring equitable and successful climate action. This entails providing women with access to knowledge, education, training, and secure land rights, as well as eliminating unfair socio-political systems that perpetuate gender disparities.

Prioritizing national capacity building is critical for climate financing projects. Investment strategies should prioritize capacity building over recovery efforts and infrastructure investment. Targeted climate finance should also be used to harness developing nations' enormous greenhouse gas abatement potential, so aiding their transition to sustainable, low-carbon economies. Tailored adaptation measures are particularly needed for Africa's agricultural business, which is extremely vulnerable to climate change, in order to promote a larger transition to a low-carbon economy.

To increase climate finance efficacy, funding must become more adaptable and risk-tolerant, while also encouraging innovation and boosting the adoption of cutting-edge technologies. Streamlining and unifying the climate funding infrastructure is critical to avoiding overlap and inefficiencies. Increasing public investment in climate finance and promoting locally driven adaptation are also critical. Finally, involving the private sector and forming new partnerships with financial institutions involved in climate-related sectors, particularly infrastructure, will generate more funding, resulting in more effective climate action.

6.2 EFFECTIVE CLIMATE FUNDING

The overriding message of these chapters is a critical shift in global climate funding, particularly for Africa and other developing countries. The recommendations create a narrative of urgent, diverse action. To effectively reduce the climate funding gap, the international community must go beyond mere pledges and deliver actual, demonstrable achievements.

For effective climate financing in Africa and other developing countries, the following actions are suggested:

1. **Transformed Policy Framework**
   First and foremost, policy frameworks must be fundamentally transformed. This involves expediting clearance procedures, creating clear standards, and encouraging climate-friendly investment. De-risking tools, such as guarantees and insurance, are critical for attracting private sector investment. This policy revamp must be accompanied by a considerable increase in financial contributions from rich countries, with an emphasis on adaptation programs and the Loss and Damage Fund, and a more fair distribution that benefits least developed countries (LDCs). A transition from loan-based to grant-based adaptation support is also required to avoid worsening existing debt loads.

2. **Capacity Building**
   Second, capacity building is crucial. Investing in training programs, technical assistance, and knowledge-sharing initiatives will enable African institutions to more effectively manage climate funding. This includes strengthening national legislation and domestic finance sources, creating strong local capital markets, and enhancing project

management skills. This requires the establishment of clear definitions and rigorous monitoring systems for tracking financial movements in order to ensure transparency and accountability. National legislation must be tightened to support aggressive emission reduction and climate resilience targets, and National Adaptation Plans (NAPs) must be prepared and executed in accordance with Climate-Resilient Green Economy (CRGE) policy.

3. **Innovative Financing Systems/Mechanisms**
   Third, innovative finance mechanisms are required. Alternative sources, such as green bonds and debt-for-nature swaps, should be investigated to diversify the resource pool and minimize dependence on traditional aid. Promoting sustainable development and renewable energy transitions is vital, including increased support for clean cooking solutions and the use of global efforts such as the Global Renewables and Energy Efficiency Pledge. The UAE FGCR should be improved with measurable adaptation success indicators, and adaptation efforts should be directed toward the most pressing concerns, particularly those in the health sector.

4. **Strategic Alliances and International Collaboration**
   Fourth, strategic alliances and international collaboration are vital. Government-to-government engagement, PPPs, and business sector participation in climate-resilient development must all be encouraged. Supporting African countries in negotiations is critical to ensuring their active participation in climate change mitigation. Overcoming obstacles such as a lack of proposal knowledge and inadequate project management necessitates targeted capacity-building programs, such as those designed after the Climate and Energy Response Facility.

5. **Paradigm Shift from Reactive Catastrophe Recovery to Proactive Capacity Building**
   Finally, a paradigm shift from reactive catastrophe recovery to proactive capacity building is required. Climate finance tools must be adapted to emerging nations' individual developmental demands, easing the transition to low-carbon, climate-resilient pathways. Modern technology such as remote sensing, GIS, and artificial intelligence should be used to increase project monitoring and transparency. Stakeholder participation, gender inclusion, and the empowerment of excluded groups are critical to equitable and effective climate action. Funding should be adaptable and risk-tolerant, promoting innovation and the use of cutting-edge technology. Streamlining the climate funding architecture, increasing public investment, and promoting private sector collaborations are critical for mobilizing resources and successfully implementing climate projects.

6.3 FUTURE PERSPECTIVES

The narrative that runs across these chapters presents a compelling picture of the future of climate finance, particularly for Africa and emerging nations. It starts with a fundamental appeal for structural change, highlighting the importance of strong policy frameworks, capacity building, and diverse funding sources. The envisioned future is one in which international collaboration is more than a talking point, with developed countries honoring their financial commitments and ethical considerations directing resource distribution. National legislation is regarded as a cornerstone, allowing countries to set ambitious goals and engage effectively in global climate debates.

As the novel progresses, the emphasis moves to a fundamental restructuring of climate finance itself. To avoid worsening debt loads, the future calls for a shift away from loan-based support and toward grant-based aid, particularly for adaptation projects. Transparency and accountability become crucial, with precise definitions, stringent monitoring, and open distribution methods ensuring that funds are spent effectively. National policies and domestic financial resources are enhanced, promoting local capital markets and prioritizing adaptation investment. The narrative predicts an increase in sustainable development and renewable energy transitions, fueled by global initiatives and clear norms for private sector participation. This future, however, is not without obstacles, demanding strong institutional frameworks to offset the negative repercussions of big financial inflows.

The narrative underlines the importance of measurable adaptation success and calls for a reinforced UAE FGCR. In the future, adaptation efforts will focus on the most pressing concerns, with increased investment on vital areas such as healthcare. Climate change considerations are incorporated into all development initiatives, as part of national adaptation plans. Quantitative indicators and robust monitoring systems measure progress, whereas adaptation initiatives focus people and address root causes of inequity. African healthcare systems are expected to receive a significant increase in financial resources, as well as the development of technological capacities for effective climate finance mobilization. To ensure openness and accountability in the future, a uniform definition of "climate finance," enhanced data collection, and the implementation of GBT are necessary.

A significant movement toward capacity building is noted, picturing a future in which African countries have the technical expertise to manage the complex climate finance landscape. Strategic alliances and collaboration become increasingly important, promoting government-to-government contact, PPPs, and corporate sector participation. International cooperation and negotiation support enable African states to actively participate in climate action rather than simply receive aid. The narrative finishes with a demand for global financial reform, emphasizing the importance of accessible and fair climate funding. The future requires a shift from pledges to tangible results, with an emphasis on solid numbers, verifiable progress, and transparency.

Finally, the narrative culminates in a vision of proactive capacity creation, rather than reactive disaster recovery. Climate finance instruments are designed to meet specific developmental demands, easing the transition to low-carbon, climate-resilient paths. Modern technologies improve project monitoring and transparency, while stakeholder involvement and gender equality are becoming increasingly important components of climate financing schemes. National capacity building takes primacy, with investment policies focusing on long-term resilience rather than speedy recovery. Funding becomes more versatile and risk-tolerant, promoting innovation and the use of cutting-edge technology. The future vision is one of streamlined climate funding infrastructure, increased public investment, and active corporate sector participation, all of which contribute to more effective and fair climate action.

Index

Note: Figures are indicated by *italics*; tables are indicated by **bold**.

For Product Safety Concerns and Information please contact our EU
representative GPSR@taylorandfrancis.com
Taylor & Francis Verlag GmbH, Kaufingerstraße 24, 80331 München, Germany